# URBAN LATINO CULTURES

The following are © 1999: Chapter 3, Don Normark; Chapter 5, text María Elena Fernández, photo Jessica Chornesky; Chapter 6, Gloria Enedina Alvarez; Chapter 7, photo Pedro Meyer, text Rubén Martínez; Chapter 8, Christina Fernandez; Chapter 9, Alma López; Chapter 10, Jesse Lerner and Rubén Ortiz-Torres; Chapter 11, Carlos Avila; Chapter 12, Lindsey Haley; Chapter 13, Luis Alfaro; Chapter 14, Lalo Alcarez; Chapter 15, Theresa Chavez; Chapter 16, text Gustavo Leclerc, photo Julie Easton; Chapter 17, Anthony Hernández; Chapter 18, Camilo José Vergara; Chapter 20, text Margaret Crawford, photos ADOBE LA; Chapter 21, photos Rita González, text Ramón García; Chapter 22, James Rojas; Chapter 23, Harry Gamboa Jr.; Chapter 24, Yvette C. Doss; Chapter 25, Rubén Martínez; Chapter 26, photos Reynaldo Rivera, text Ramona Ortega; Chapter 27, Rogelio Villarreal Macías; Chapter 28, John A. Loomis; Chapter 29, Robert Alexander González.

*For information:*

SAGE Publications, Inc.
2455 Teller Road
Thousand Oaks, California 91320
E-mail: order@sagepub.com

SAGE Publications Ltd.
6 Bonhill Street
London EC2A 4PU
United Kingdom

SAGE Publications India Pvt. Ltd.
M-32 Market
Greater Kailash I
New Delhi 110 048 India

Printed in the United States of America

*Library of Congress Cataloging-in-Publication Data*

Main entry under title:

Urban Latino cultures: La vida latina en L.A. / edited by
Gustavo Leclerc, Raúl Villa, and Michael J. Dear.

    p.  cm.
  Includes bibliographical references and index.
  ISBN 0-7619-1619-9 (cloth: acid-free paper)
  ISBN 0-7619-1620-2 (pbk.: acid-free paper)
  1. Hispanic Americans—California—Los Angeles.  2. Los Angeles
(Calif.)—Civilization—20th Century.  I. Leclerc, Gustavo.
II. Villa, Raúl.  III. Dear, M. J. (Michael J.)
  F869.L89 S758    1999
  979.4′9400468073—dc21           98-58087

99  00  01  02  03  10  9  8  7  6  5  4  3  2  1

| | |
|---|---|
| *Acquiring Editor:* | Catherine Rossbach |
| *Editorial Assistant:* | Heidi Van Middlesworth |
| *Production Editor:* | Diana E. Axelsen |
| *Editorial Assistants:* | Karen Wiley |
| | Nevair Kabakian |
| *Permissions Editor:* | Jennifer Maxon Morgan |
| *Interior Designer:* | Ravi Balasuriya |
| *Typesetter:* | Marion Warren |
| *Cover Designer:* | Ravi Balasuriya |

LA VIDA LATINA EN L.A.

# URBAN LATINO CULTURES

EDITED BY

## GUSTAVO LECLERC
## RAUL VILLA
## MICHAEL J. DEAR

PUBLISHED IN ASSOCIATION WITH THE
SOUTHERN CALIFORNIA STUDIES CENTER
OF THE UNIVERSITY OF SOUTHERN CALIFORNIA

**SAGE** Publications
*International Educational and Professional Publisher*
Thousand Oaks   London   New Delhi

## Photos of Everyday Life

We would like to extend our special thanks to the people who permitted us to use their photographs of everyday life in Los Angeles. These occur throughout the text without captions and are used by permission of the following:

| | |
|---|---|
| Page 1: | Julie Easton, ADOBE LA |
| Page 2: | Michael Dear |
| Page 3, top: | Ulises Diaz, ADOBE LA |
| Page 3, bottom: | Julie Easton, ADOBE LA |
| Page 4, top: | Julie Easton, ADOBE LA |
| Page 4, bottom: | Ulises Diaz, ADOBE LA |
| Pages 7, 8: | Gustavo Leclerc, ADOBE LA |
| Page 9, top: | Social and Public Art Resource Center (SPARC), Venice, California |
| Page 9, bottom: | Gustavo Leclerc, ADOBE LA |
| Page 11: | Julie Easton, ADOBE LA |
| Page 12: | Ulises Diaz, ADOBE LA |
| Pages 14, 15: | Ulises Diaz, ADOBE LA |
| Page 16: | Julie Easton, ADOBE LA |
| Page 28: | Julie Easton, ADOBE LA |
| Page 144, top and bottom: | Ulises Diaz, ADOBE LA |
| Pages 145: | Julie Easton, ADOBE LA |
| Pages 146, 147: | Ulises Diaz, ADOBE LA |
| Pages 148, 149, 150, 151, 152, 153: | Julie Easton, ADOBE LA |
| Page 154, top: | Alessandra Moctezuma, ADOBE LA |
| Page 154, bottom: | Ulises Diaz, ADOBE LA |
| Page 156: | Julie Easton, ADOBE LA |
| Pages 157, 158-159, 160, 162-163, 170, 172, 173, 174: | Raúl Villa |
| Page 185: | Michael Dear |
| Pages 186, 187: | Julie Easton, ADOBE LA |
| Page 188: | Michael Dear |
| Pages 189, 190, 191, 192: | Julie Easton, ADOBE LA |
| Pages 194, 196: | Michael Dear |
| Page 197: | Jennifer Wolch |

# Contents

# Agradecimientos/
# Acknowledgements

Our greatest appreciation goes to our contributors, whose works made this book possible.

Special thanks also to Ulises Diaz of ADOBE LA for his many contributions to the book's design, and for his advice and support.

We are grateful for the financial support of the Southern California Studies Center (SC2) of the University of Southern California through its program on Race, Ethnicity, and Place. The Center is funded by The James Irvine Foundation, and by the University of Southern California through the offices of President Steven Sample, Provost Lloyd Armstrong Jr., and Dean of the College of Letters, Arts and Sciences, Morton Schapiro. In the SC2 office, Dallas Dishman, Becky Montaño, Lawrence Mull, Richard Parks, and Heidi Sommer were always on hand to help.

Various parts of our project were also supported by Occidental College, the Southern California Institute for Architecture (SCI-Arc), the Cultural Affairs Department of the City of Los Angeles, and by The John Randolph Haynes and Dora Haynes Foundation. Thanks to Neil Denari, Michael Rotondi, and Margaret Crawford at SCI-Arc.

Gustavo offers special thanks to Barbara Jones.

Once again, it is a great pleasure to acknowledge our friends at Sage Publications. Editor Catherine Rossbach recognized the potential of this volume even before we did. And Diana Axelsen's team at Thousand Oaks expeditiously and expertly guided us through the production process. Thanks to Diana Axelsen, Ravi Balasuriya, Alison Binder, Nevair Kabakian, Michèle Lingre, Heidi Van Middlesworth, Jennifer Morgan, Marion Warren, and Karen Wiley. Working in three languages (Spanish, English, and Spanglish) was never a problem for this cool crowd.

¡A todos, muchas gracias!

# CHAPTER 1

# Introduction:
# La vida latina en L.A.

Gustavo Leclerc
Michael J. Dear

This book is about a cultural revolution that is occurring now in Los Angeles and other great cities of the United States. On the streets, in the media, inside their homes, and in public places everywhere, Latinos are reclaiming and re-creating their cultural heritages. Latinos are now the majority population in many cities of Southern California; their numbers have increased because of immigration; and they are voicing a desire to become equal partners in social, political, and economic institutions in this country. Such challenges pose acute problems for other ethnic and racial communities but also for Latinos themselves, who are faced with increasing diversity within their ranks and a consequent divergence of objectives. There can be no doubt that far-reaching changes are occurring in the demographic makeup of Southern California and, with them, important adjustments in the way people relate to one another. This book is a document of these changes. It collects evidence of a revolution as manifest in cultural events on the street; in magazines, art, and television; and in universities, homes, and the workplace.

In 1848, through the Treaty of Guadalupe Hidalgo, the territory known as *Alta California* became part of the United States, erasing the legacies of Spanish, Mexican, and Indian traditions. Except, of course, these traditions never went away—they persist

as palimpsests undergirding everything in contemporary Southern California. They are visible, for example, in settlement patterns, the names of streets, and the way L.A. boulevards follow the boundaries of former *ranchos.*

There is no more potent symbol of L.A.'s Latino geography than the canyon of concrete walls and seasonal runoff known as the L.A. River. To white Angelenos, the river represents the displacement of nature by urban growth, but for Latinos, it is a manifestation of the separation between East and West, the dominant (white) culture and marginalized (nonwhite) cultures. It is a physical expression of all the internalized Latino/Chicano/Mestizo metaphors of *difference.* Exiled beyond the river, Latinos are finding new ways to (re)create their own identities. Such "cultures of resistance" can take many forms, but all have to do with people and *place.*

It is often said that *place makes a difference,* that the place where you grew up somehow leaves an indelible mark on your character and outlook. You experience this every time someone asks where you come from, or what's that accent you have. But it is equally true that *difference makes places,* in the sense that diverse people create recognizably different spaces for themselves. For instance, you know when you've crossed into the *barrio* simply by looking at house and garden decorations, street and store signs, people gathering on sidewalks and in parks. This book is not only a document of an emerging cultural revolution but also a meditation—on the difference that place makes and the places that difference makes.

Now, how place is created and what places enable us to do are two complicated issues that lie at the heart of the concerns of geographers and urbanists, architects and planners, and artists and other cultural producers. We have simplified our task in this book by focusing on one city, Los Angeles, and especially the *barrios* of East L.A. Also, we have concentrated our attention on the cultures of everyday life, as recorded by artists and *barrio* residents themselves. But we record perspectives from other cities as well, including New York and Mexico City, in the belief that what is happening in Southern California is typical of many cities across the United States.

We chose Los Angeles because of the size, variety, and growing significance of its Latino populations but also because Southern California is regarded by many as the prototype of twenty-first-century urban growth. The region is likely soon to surpass New York as the nation's largest metropolis. It is already America's preeminent focus on the emerging Pacific Rim and is the site of some of the most significant sociocultural, political, and economic

innovations.

We chose the cultures of everyday life because we believe that some of the keenest investigations into our emerging social compact (i.e., the terms under which people agree to live together in relative harmony) are being undertaken by artists, as well as scholars, planners, architects, and other urban professionals. Many of the most vital visions of our altered urban future are to be found at the street level, in homes, cars, clothing, games, art, and so forth, that is, in the cultural lives of ordinary people.

But questions remain. How are these artistic investigations and cultural expressions formed? From what are urban lives and representations created?

*The creation of contemporary Latino identity emerges from the intersection of memory, person, and cultural mixing.* More specifically, the dimensions of this process are these:

*Memory and the South,* referring to the primordial origins of ancestry, identity, tradition, and selfhood;

*Body and Identity,* referring to the individual (body and mind) in the dynamic evolution of identity formation; and

*Borders, Hybrids, and Resistance,* referring to the volatile synergies of old and new places (e.g., ancestral home, border crossing, and *barrio* exile) in identity, participation, and resistance.

The works collected in this book portray the outcomes of the dynamic interaction among these various forces. Many contributions incorporate multiple themes and do not fit readily into a single category; the dimensions persist, however, as threads to guide the reader. They are, in essence, the keys to the cultural cartographies excavated in this book. It is important, therefore, that we spend a little more time explaining them.

*Memory, or Remembered Space.* Memory deals with the history of Los Angeles, a constructed past that has excised our knowledge of entire communities. Much of L.A.'s Latino heritage has been lost to the city's official history, creating a void in the construction of Latino selfhood. Yet cities, through their urban and architectural forms, are also unconscious records of memory. By sifting through the narratives of buildings, streets, and neighborhoods, it is possible to glimpse a complex set of histories and geographies that can lead to new understanding of the formation of urban identity.

*South, or Chimerical Space.* The South is being simultaneously created and re-created in Mexico and Los Angeles. Alternative ways of producing cityscapes are being negotiated, and original ways of seeing and speaking about these emergent spaces are being invented. This alternative discourse is not based in or legitimized by European or colonialist concepts. Instead, it is being distilled in the spaces between dreams and longing, somewhere between L.A. and the South. Its locus is *el norte, el otro lado,* even as it remains firmly embedded in the myths and traditions of the old country. Here lies another palimpsest (a trace, footprint, or chimera) that cannot be erased even as it is transformed.

*Body, or Performative Space.* From origins in remembered and chimerical spaces, personal consciousness is formed. Performative space refers to a body's response to being in specific places; it refers to how we act in a (new) place. For individuals, the body is the ultimate site of resistance to social and economic controls; it is the site of collective and emotional linkages as well as personal vulnerability. Latino bodies are both sensuous and critical, assimilative yet assertive, displaced yet rooted, as they encounter the transformations of their everyday lives.

*Identity, or Experienced/Lived Space.* Taken together, the collective experiences of countless individuals define identity and meaning in a particular place, along lines of (for instance) class, gender, and ethnicity. In a reciprocal manner, the qualities of place

act to condition and constrain the mechanisms of identity formation; thus, the neighborhood protects us, at the same time as it holds us back, keeps us down. One of the most pervasive places in Latino identity is the border, with its constant possibility always for crossing . . . and re-crossing . . .

*Border Crossing, or Liminal Space.* Borders affect identity because so many Latinos are part of monumental migrations bridging notions of past and present, home and dislocation. In the between-spaces, or *liminal spaces,* new *hybrids* are concocted whereby elements of the different worlds simultaneously coexist. And, as nation-state borders become more permeable, a transnational "postborder" condition complicates the formation of local identity. This postborder condition is defined by a simultaneous centrality and marginality, by ambiguity and ambivalence. It is creating in Southern California a new type of liminal space—a transnational borderland, a type of blurred macrofrontier.

*Hybridity, or Transformative Space.* Hybrid social spaces result when juxtapositions occur between the First and Third Worlds, between rich and poor, between North and South, between local and global, between modernity and postmodernity, and between Spanish and English. In such spaces, cultural production springs from the tensions between indigenous and migrant cultures. Hybrid spaces are coinhabited by individuals and groups of multiple origins, who adjust to the city around them just as they are changing the city itself.

*Resistance, or Appropriated Space.* As in other cities, powerful interest groups have controlled access to public spaces in Los Angeles. Their concerted efforts have kept "undesirable" people and "unwanted" uses away from public places, certain buildings, and whole neighborhoods. But the appropriation and reclamation of spaces by marginalized groups can make places effective sites of urban intervention and sociopolitical opportunism. This is often expressed in the forms of insurgent urbanism and architecture, including the painting of murals or the practice of tagging, which usurp the totalizing discipline of the city's built forms.

In summary, the cartographies of these seven spaces together define the principal processes (memory, the dynamics of identity creation, and hybrid geographies) that are producing the cultural changes documented in this book.

■

*Urban Latino Cultures: La vida latina en L.A.* began with an idea of creating a cross-disciplinary forum to examine issues of Latino identity and representation in Los Angeles. Our goal was to open up a new space to discover the construction of individual and civic cultural identities in Latino communities. In our dreams, we wanted this space to be multilingual, panethnic, and transnational. Accordingly, the individuals who have contributed to this book come from a variety of places and backgrounds, ranging from scholars to performance artists to literary critics. Most live in Southern California, but almost all have come from other places. Some have put down roots in other cities such as New York and Mexico City. Most contributors are Latino, but in their work, every one confronts issues of Latino identity and the production of urban social space. They speak Spanish, English, and Spanglish—sometimes all three. Their contributions encompass multiple forms of cultural criticism, from narrative histories to intimate portraits, from documentary photography to in-your-face satire, from rock opera to sweatshops.

Our collection comes at a critical time in California's political history, when tension among racial and ethnic groups is escalating. This is partly the result of the larger borderland relationships between the United States and Mexico and much of Latin America, which are rife with contradiction and conflict. For instance, although NAFTA brought fresh economic opportunities, it also aggravated social inequalities, cultural misunderstandings, and mistrust. And yet, to this point, there has been little constructive dialogue around questions of representation, popular culture, and urban interdependency between Latin America and the United States. Because of these interdependencies, millions of Mexican/Latino immigrants are living within the United States today. As cities begin to physically express the cultural diversity that is their inheritance, there have been some negative reactions. California's unease is recorded in the rush of voter initiatives including Propositions 187 (against state services for undocumented immigrants), 209 (against affirmative action), and 227 (against bilingual education). But there can be no turning back, no turning away from the challenges of coexistence. To borrow words from Raúl Villa: "Aquí estamos y no nos vamos" "We're here, and we're not going away."

Times of the greatest challenge are also moments for opportunity and creativity. As we write, a great social experiment is occurring as unprecedented varieties of peoples invent new ways to live together. We have little doubt that a new "social contract" is currently being forged in Los Angeles. No one knows how this experiment will turn out or what this new social contract will look like. In the pages of this book, we have attempted to record one pivotal moment in this tumultuous evolution. And we do this with the full conviction that something enduring and wonderful is occurring now, right in front of our eyes, in what used to be called *Alta California*.

# CHAPTER 2

# Aquí estamos y no nos vamos
# Place Struggles in Latino Los Angeles

Raúl Villa

Place expresses how a spatially connected group of people mediate the demands of cultural identity, state power, and capital accumulation.

*Sharon Zukin*[1]

On May 16, 1877, the editors of *La Crónica*, a Spanish-language newspaper in Los Angeles, railed in print against the systematic dislocation of the Mexican population from its once-central position in the cultural landscape of *El Pueblo de Nuestra Señora la Reina de Porciúncula de Los Angeles:* "We still have a voice, tenacity and rights; we have not yet retired to the land of the dead." Just over a hundred years later, Harry Gamboa Jr., multimedia artist *provocateur*, argued that in the dominant cultural view, "We [Latinos] are seen as a phantom culture" and used the creative resources at his disposal to combat this mainstream projection.

Separated by time but linked by intention, both statements note how the powerful mechanisms of a dominant society effectively "disappear" Latinos from view, relegating them to the shadows of the Anglo imagination and metropolitan social order. Although the recurring need to speak out against the Latino community's material and ideological displacement suggests the

efficacy of Anglo capitalist urbanization, it also calls attention to the historical trajectory of Latinos defending their place and presence in the region. Echoing across the span of Los Angeles's modern history—from its preindustrial origins as an early Anglo township to its post-Fordist emergence as a global megalopolis—these twin declarations serve as discursive quotations marks for a cumulative position statement of Latino Angelenos: "Aquí estamos y no nos vamos."[2]

If many Latinos in contemporary Southern California, like their Californio predecessors, have not yet retired to the land of the dead, it is not for lack of external pressures to do so. The disparate impacts of capitalist urban development have posed real material threats to the well-being of poor and working-class barrio dwellers. The continuous restructuring of the central city has regularly disrupted their community lifeways, erased or exoticized their cultural landscapes, and threatened the spatial basis of their collective political organization. In the process, Latino community spaces have often been "developed" for purposes quite removed from the desires of their residents.

In their continued desire to realize Los Angeles's Anglo American manifest destiny, civic elites, private developers, and urban planners have coordinated their efforts in successive projects of monumental city building: from turn-of-the-century City Beautiful design, through the rational exercises of "higher and better" land use in 1950s and 1960s urban renewal and freeway construction, to the trickle-down economics of contemporary redevelopment. Each of these periods of metropolitan renovation, however, has been achieved at disproportionate expense of barrios and other poor inner-city communities, whose residential milieus often fall directly in the path of physical urban restructuring.

The relationship of Latino residents to the machinations of Los Angeles's urban growth thus manifests the creative destruction characteristic of capitalist urbanism generally, in which, to draw from Marx and Engels's *Communist Manifesto*, "All that is solid melts into air." Translated into urban planning terms, the capitalist imperative of increasing surplus accumulation (i.e., profit) compels the constant transformation of metropolitan real estate toward "higher and better" uses. In this context, the social, cultural, and economic uses that barrio residents, like the inner-city working class generally, make of their urban milieus are invariably poorer in generating profits and tax revenues, the *raison d'être* of the city organized as an economic growth machine. For Mexicans in particular (as the oldest and most numerous of the various Latin American national origin communities in Los Angeles, as well as its largest "minority" population), their contradictory social location—being simultaneously in the geographic center *and* the economic margins of the city—has meant that they are constantly having to react to the disparate impacts of metropolitan restructuring in defense of their urban needs.

During the first major transformations of the built environment in the 1870s and 1880s, the Spanish-language press, as I suggested in opening, took the lead in condemning the discriminatory allocation of urban services and infrastructural construction in the developing city. In this capacity, they engaged in the first explicit critiques of repressive "barrioization" as it was beginning to structure the subordinate social and spatial location of Mexicans in the urban order. Recognizing the symbolic value of the Mexican built environment, José Rodríguez, editor of the weekly *El Joven,* pointedly objected to the planned demolition of Pio Pico's home in the vicinity of the central plaza, the historic heart of the pueblo. Rodríguez noted its significance as a land-

mark of Mexican history in the region, since Pico—the last Mexican governor of the region and a citizen of great standing in the Mexican period—had used his house as a seat of governance where Californio leaders met to deliberate issues of importance to the region. Although the house was eventually razed, the protest in print was one of many interventions by which Mexicans articulated and defended their place rights in the initial period of Los Angeles's aggressive Anglo-dominated expansion.

The next major period of urban metamorphosis in the 1910s and 1920s was dramatically signed in space by the expansion of downtown railroad and warehouse facilities and the construction of the new civic center complex of government and commercial buildings. In this reconstruction, the remaining barrios adjacent to the plaza found themselves squeezed out to accommodate the growth. Thousands of Mexicans were consequently forced to relocate to the next ring of central city land and housing. In the process, they seeded the ground of several important barrios in the immediate north, east, and west of downtown, including Lincoln Heights, Chávez Ravine, Boyle Heights, and Bunker Hill. The instrumental practice of large-scale urban removal, under the guise of "urban renewal," was thus first fully exercised on the residential places of working-class immigrants in the central city. This included not only the growing Mexican community, fueled by the push factor of the Mexican Revolution, but also significant populations of Chinese, Italians, and Eastern Europeans, among other groups.

Although the Anglo American leadership expressed its rule in the monumental transformations of downtown, a different practice of urban signage came into being. Mexican youth gang graffiti emerged in the shadows of central city expansion during the 1930s, perhaps even in the 1920s. Appearing in marginal spaces of the new downtown and surrounding barrios in which many uprooted Mexicans resettled, these other marks on the landscape signaled a distinct relationship to urban space for those citizens pushed aside in the growth of the "wonder city." This early turf tagging by young Mexicans expressed a need to claim and personalize urban places that was compelled by the exacerbation of affordable housing and public space for Mexicans. Although the now familiar public outcry for graffiti abatement and graffiti-busting legislation would not surface for many decades, mainstream Anglo American consciousness of a nascent but rising gang presence was piqued from the beginning by Emory Bogardus's groundbreaking studies of Mexican social problems.

Bogardus's University of Southern California students methodically canvassed the inner-city slums, in the process compiling a cumulative "scientific" portrait of Mexican maladjustment, particularly in the form of youth gang deviancy. The USC School of Urban Studies gave scholarly support to the Los Angeles Police Department's nascent statistical profile of Mexican criminality. In this period, the very use of public space by Mexicans became a constituent category of their alleged social and criminal maladjustment. For example, a 1930 state fact-finding commission documented "vagrancy" as a leading cause of arrest for Mexican men, second only to drug violations. This fact is not surprising because antivagrancy "greaser laws" were first enacted by the city's common council in the 1870s. The commission report also listed loitering as the principal reason for police citation of Mexican juveniles. The current attempt by the district attorney's office to curb youth gang activities by restricting the right of congregation on public streets and parks is clearly not a novel approach to this long-standing urban phenomenon.

Even as scholars, social reformers, and the police were attempting to analyze and regulate the activities of the perceived aliens in their midst, however, a growing number of Mexicans were accommodating themselves to their new urban environment in ways that signaled a collective desire to establish their place as Mexicans *de este lado*. A conspicuous measure of this place-making process was the steady increase in home ownership by immigrant Mexicans, who chose to settle, start families, and raise their increasingly American-born children in the United States. Occurring primarily in cheaper outlying and unincorporated county land east of the Los Angeles River, the expansion of single-family resident home ownership in areas such as Maravilla and Belvedere during the 1920s and 1930s marked the birth of the East

L.A. superbarrio to come. This trend also debunked the myth of Mexican labor transience popular in the Anglo American imagination of the time. The myth held that Mexicans did not really need to be reckoned with because an innate "homing" instinct would invariably draw them back to their native land in Mexico.

The place-rooting intentions of Mexicans were even more conspicuously manifest in the local defense of Belvedere against several incorporation attempts by the city of Los Angeles. Neighborhood residents were informed in their actions by a clear consciousness of external threat, drawing on their knowledge—for many, from direct experience—of previous community displacements in the central city. Thus, Belvedere residents saw the incorporation plans as a thinly masked ploy by the city council, acting in concert with private developers, to pressure the Mexicans out and gentrify the annexed land. The failure of this early growth coalition to take over Belvedere proved the will of this nascent Mexican working-class community to develop and sustain its place in the greater urban map.

The major urban developments enacted in the early twentieth century were a rehearsal for the much greater exercise of repressive urban planning on the integrity of Mexican American and eventually greater Latino milieus in the city after World War II. Already in the early 1940s, the violent transgressions of Mexican social spaces during the misnamed "Zoot Suit Riots" gave dramatic public notice to Chicanos of their vilified location in the urban order. In the conflagrations of June 1943, U.S. servicemen invaded public and private places of Chicano congregation throughout the city—including principal streets and boulevards, movie theaters, streetcars, homes, and restaurants—to harass, beat, and strip young men of their sartorial and "un-American" zoot suits. These assaults were often conducted with the help or tacit approval of local police agencies. Several major newspapers exhorted "patriotic" Angelenos to attack this perceived threat to civic well-being and the greater war effort, thus making the young Chicanos into folk devils against whom the "good" citizens exorcised their own wartime anxieties. In like fashion, although not to the same extent, Chicanos engaged in their own defensive guerrilla strategies to protect themselves and their barrios, setting out ambushes for unwitting servicemen who wandered into their turf. Barrio folklore documents numerous accounts of such retaliatory efforts by young men, and it has been a theme in literary and autobiographical narratives of growing up Chicano in this period.

In a broad sense, the assaults on young Chicanos and their social spaces during World War II were prelude to much larger-

scale violations of their community's residential milieus yet to come. The Eastside, and other compass points in the immediate circumference of downtown Los Angeles, would be aggressively altered by large-scale urban highway construction and urban renewal projects beginning in the 1950s and 1960s. Rodolfo Acuña, the major historian of these infrastructural impacts on the barrios, convincingly argues that they amounted to a "state of siege" on the greater Mexican-Latino communities in the central city and especially East L.A. Two of the most spectacular instances of spatial violation against Mexicans and other poor people in the central city were the displacement of the barrios in Chávez Ravine to the north for the construction of Dodger Stadium and the vivisection of Boyle Heights and the greater Eastside barrios to make way for the East L.A. freeway interchange and the several highways that radiated out from it.

These signal developments of the 1950s and 1960s materially facilitated Los Angeles's next transformation into the nation's supercity and symbolically represented this image to the outside world. In the process of their realization, the arguments of Latino residents for how to best use their neighborhoods were dismissed as parochial, if not wholly regressive, by the urban elites spearheading this growth. As in previous cycles of major development, barrio-based concerns proved of little consequence. Economistic analyses "proved" the greater metropolitan benefit of the projects slated for the prime downtown land occupied by the barrios.

Cloaked in an aura of high modernist civic purpose, city boosters asked rhetorical questions that supported their plans. What great American city could be without a professional baseball franchise? What could outweigh the need to provide swift and safe automotive access to and from downtown for the growing suburban workforce? Then Mayor Norris Poulson took the modernizing rhetoric a step further. In a public radio address to the city in 1959, he chastised the critics of his administration's growth plans: "If you want Los Angeles to revert to pueblo status, if you want nothing changed . . . then my best advice to you is to prepare to settle elsewhere, because whatever you may do . . . cannot stop the momentum which is thundering this city to greatness."[3]

Although the worn but familiar environs of the Chávez Ravine and Boyle Heights neighborhoods provided an essential range of social, cultural, and economic resources for their residents, these could not measure up against the spectacular future projected in the city's plans for these areas.

Despite the grandiose arguments of urban elites, however, Eastside residents living in the path of the growth machinery rallied to combat the "bulldozers in the barrios," to borrow Rodolfo Acuña's pithy characterization of postwar freeway construction and urban renewal. Their protests against the siting of freeways in their neighborhoods took the form of public rallies, appeals to their elected representatives, and editorial interventions in the community weekly, the *Eastside Sun.* The imperious use of eminent domain by local and federal officials in forcing the hand of local property owners was a fundamental issue of contention. But although many residents objected at root to the building of freeways in the community, many understood the inevitability of their construction. In light of this realization, their objections concerned the poor compensation for loss of property, the flawed provisions for relocation, and the inadequate subsidies and consideration for the plight of renters with the net loss of good, affordable housing stock in the central city. Ironically, after initially protesting the coming of the freeways, Eastsiders subsequently had to rally to have freeway exits *put into* their neighborhoods. The bias of the planners was revealed in this neglect of local access to the Eastside, which would have been cut off from the greater metropolitan mobility promised to suburban commuters.

Although they lacked the political power to stop the assault on their milieus by freeway construction, unlike the more affluent communities of Beverly Hills and South Pasadena, their struggles revealed that Eastsiders were not without will or agency in arguing for their urban needs. The battles around the redevelopment of Chávez Ravine showed similar volition on the part of barrio residents. The net impact of freeway construction—involving such things as land loss, residential displacement, environmental degradation, and general disfigurement of the Eastside community's social geographic nexus—far surpassed the effects of the Chávez Ravine land grab, but the symbolic registers of the latter were much greater. Although the building of freeways as a public works project was arguably necessary for the efficient running of the city, the turn of events in Chávez Ravine revealed a more naked relationship between private enterprise (in the form of Walter O'Malley's Dodgers) and public power.

For several decades, the hilly landscape of Chávez Ravine had been home to those unable to procure housing elsewhere. Over-

whelmingly Mexican in population, the Ravine was also home
to African American and Chinese families as well as single,
indigent Anglo men. The area
was terra incognita to the majority of Angelenos and was
effectively left to its own de-
vices by city officials. Although
it sustained a church, an
elementary school, and a local
newspaper, the real lifeblood of

Chávez Ravine was its interpersonal networks of support. These
everyday networks facilitated the exchanges of goods, services,
and information on which the residents, like barrios dwellers past
and present, depended for their material and cultural well-being.
These included such things as child care, job leads, home medical treatment (*curanderismo*), bartering for goods and services,
money loaning, and miscellaneous activities tied to bonds of
friendship and extended-family relations. These forms of social
capital helped compensate for the lack of economic capital
among residents, while bringing them into a social compact
with one another wherein personal obligation for repayment of
services or goods was more than a contractual agreement. These
relationships, compelled and cemented by the exigencies of survival among the poor, produced a truly felt sense of community
among residents of Chávez Ravine.

The relative isolation of these barrios and their intricately wired
social networks would be shattered, however, after the area was
selected by the city's housing authority in 1949 as the site for its
most ambitious public housing project. Although there was some
opposition to the city's plans, the assurances given to residents
about the viability of the project as well as the promise that they
would be given priority in allotment of new housing allayed many
of their fears and allowed the housing authority to purchase land
(with federal subsidy) in the necessary quantity. For once, it
seemed that the benefits of urban modernization would be dis-
tributed to a class of citizens who typically only bore its costs. It
was a short-lived fantasy.

Riding the crest of a fierce red-baiting campaign that maligned
public housing as socialistic, downtown real estate interests were
able to elect a puppet candidate, Norris Poulson, in 1953 over
incumbent mayor Fletcher Bowron. Poulson quickly scrapped
the plans for new housing, allowing the acquired land in the Ravine to lay fallow while a suitable public purpose, stipulated by

law, was being determined. That purpose turned out to be luring the Brooklyn Dodgers west with a gift of the land (technically an "exchange" of the 315-acre Ravine parcel for the 9-acre Wrigley Field site in South Los Angeles belonging to Dodgers' owner Walter O'Malley) and the promise of $2 million in publicly financed infrastructural development. The looming specter of communism evoked to defeat subsidized housing for the poor became a friendly ghost when offering O'Malley tremendous material incentives to bring the boys of summer to the land of sunshine.

Former residents and other citizens joined those few families remaining in the Ravine to protest the double standard of welfare for the rich and fight against the coming of the Dodgers. The battle of Chávez Ravine was waged with great fervor by both sides and with full media coverage. The level of engagement led Mayor Poulson to describe it in retrospect as "the hottest battle in California since the war with Mexico."[4] The racialized implications of this turf war were clear to him and to the displaced residents in their losing cause. To this day, former residents of the area meet for an annual picnic in Elysian Park, adjacent to the buried neighborhoods, to commemorate the tragic fate of their old community. Although this anniversary event of *Los Desterrados,* as they refer to themselves, clearly acknowledges the powerlessness of the working-poor Mexican community against the public-private growth machinery, the ritual remembrance nonetheless signals their refusal to let the inequities of history be shadowed and forgotten. In this respect, their actions echo the declaration of the 1877 Californio editorial that spoke for an earlier community of *desterrados* that refused to be relegated to the shadows of the dominant metropolitan imagination.

Although the odds have long been stacked against the needs and wishes of barrio residents, they have gleaned practical knowledge from the trajectory of urban tragedy and struggle. If the 1950s and 1960s witnessed the devastations of large-scale physical restructuring in the barrios, community-based groups in the 1970s, such as the East/Northeast Committee to Stop Home Destruction, drew on these historical lessons to formulate political strategies that halted or modified the city's urban renewal plans in the Pico-Union, Lincoln Heights, and Temple-Beaudry neighborhoods. Similarly, the community defense organization Mothers of East L.A. (MELA) turned personal tragedy into political consciousness. Several of the group's founding members lost homes to freeway construction in the 1950s, whereas others had friends and family who experienced displacement. Learning from their losses, MELA members rallied in the 1980s and 1990s to

defeat several plans for siting environmentally hazardous industrial facilities and prisons on the Eastside.

The battle continues to be waged between the imperial designs of downtown growth interests and the daily needs of poor residents of color in the circumferential barrios and ghettos of the central city. The newest monuments to Los Angeles's emergence as a world metropolis continue to go up in the corporate and cultural citadel of Bunker Hill, site of the last remaining downtown barrios that were razed en masse in the 1960s. This centralized enclave of global finance and high art offers the most recent and egregious example of welfare for the wealthy administered at the expense of the poor. Responding to the peak of violent and peaceful civil unrest in the inner city in the late 1960s and 1970s, city leaders promised to invest public funds into redeveloping the Latino and black neighborhoods of East and South Central Los Angeles. Expectations were raised when Tom Bradley, an African American council member from South Central, was elected mayor in 1973 on a platform of progressive urban reform. His victory was secured with the support of both inner-city residents and liberal Westsiders. During his three terms in office, however, Bradley turned his back on the inner-city communities that helped elect him, diverting necessary funds from the barrios and ghettos into the trophy-building projects downtown. The trickle-down benefits projected for inner-city residents in the rebuilding of Bunker Hill were dammed up somewhere along their way. In this zero-sum redevelopment shuffle, then, the rise of the city's new acropolis pushed East L.A. and South Central (the latter now overwhelmingly Latino) further down in the city's socioeconomic pecking order.

Los Angeles barrio residents now represent all the Central American and several Caribbean and South American nationalities. Although the nativist legislation of California Propositions 187, 209, and 227 pose added threats to the social place of Latinos and other ethnic communities with large immigrant constituencies, this variegated *raza* carries on its place-making and place-rooting activities. These run the gamut from the localized sphere of people's daily rounds to the broader public sphere of the city and state. A sampling of these practices includes the following, among many others: reviving declining "main street" commercial strips with culturally specific enterprises, as on Pacific Boulevard in Huntington Park or the downtown Broadway corridor; practicing electoral politics, as seen in the 1996 election of Loretta Sánchez to Congress in Republican Robert Dornan's former Orange County district and her reelection in 1998; the increasing Latino voter registration in the central city; expanding the number and scope of community service organizations, such as the Esperanza Community Housing Corporation in South Central and the Central American Resource and Education Center in Westlake/Pico-Union; reinvigorating the labor movement through the activism of service unions such as Justice for Janitors; and simply enlivening public space through the intensive use of parks and derelict lots for family gatherings, soccer games, and informal commerce. In these and many other ways, some of them addressed elsewhere in this book, Latinos stake their claims to cultural, if not always national, citizenship and remake the city from the ground up in their own image.

The impulse among Latinos in Los Angeles to create and enrich their place in the city is ever engaged in a tactical war of positions with top-down urban plans and ideologies. My point, therefore, is to draw attention to a fundamental fact of metropolitan life and civic culture: that the city is a meeting ground of contending social forces and contentious wills to being-in-place. To foreground this urban dialectic is not to underestimate the predominant capacity of ruling groups or classes to shape the broad contours of L.A.'s social geography, including the place of Latinos in it. But it must also be stressed that the forms, functions, and cultural meanings of urbanity are produced simultaneously, and differently, by working-class Latino residents. Like other immigrant communities, they have contributed greatly, if not always in acknowledged ways, to Los Angeles's storied growth.

## Notes

1. Sharon Zukin, *Landscapes of Power: From Detroit to Disney World* (Berkeley: University of California Press, 1991), 12.

2. This chapter draws on several important studies for historical information and for theoretical orientation toward the topic. These are: Rodolfo Acuña, *A Community Under Siege: A Chronicle of Chicanos East of the Los Angeles River, 1945-1975* (Los Angeles: Chicano Studies Research Center, UCLA, Monograph No. 11, 1984); Michel de Certeau, *The Practice of Everyday Life* (Berkeley: University of California Press, 1984); Richard Griswold del Castillo, *The Los Angeles Barrio, 1850-1890: A Social History* (Berkeley, University of California Press, 1979); Henri Lefebvre, *The Production of Space,* trans. Donald Nicholson-Smith (Oxford, England; Cambridge, Mass.: Blackwell, 1991); John R. Logan and Harvey Molotch, *Urban Fortunes: The Political Economy of Place* (Berkeley: University of California Press, 1987); Carey McWilliams, *North from Mexico: The Spanish-Speaking People of the United States* (New York: J. B. Lippincott, 1948); Mary Pardo, "Identity and Resistance: Mexican American Women and Grassroots Activism in Two Los Angeles Communities" (dissertation, University of California, Los Angeles, 1990); Don Parson, "The Development of Redevelopment: Public Housing and Urban Renewal in Los Angeles," *International Journal of Urban and Regional Research* 6, no. 7 (September 1982): 393-413; Jerry Romotsky and Sally R. Romotsky, *Los Angeles Barrio Calligraphy* (Los Angeles: Dawson's Book Shop, 1976); George J. Sánchez, *Becoming Mexican American: Ethnicity, Culture and Identity in Chicano Los Angeles, 1900-1945* (New York: Oxford University Press, 1993).

3. Quoted in Joseph Eli Kovner, "East Welfare Planning Council Is Tool of Mayor's Urban Renewal Committee, as Is West LA Group!" *Eastside Sun,* 8 January 1959, p. 1.

4. Parson, "The Development of Redevelopment," 403.

# CHAPTER 3

# Chávez Ravine

Don Normark

On a clear day in December 1948, while looking for a high place to photograph in Los Angeles, I found what seemed to me a hidden village. The place was Chávez Ravine, and the village was called *La Loma* by the Mexican Americans who lived there. The people I en-countered were friendly, in-cluding those elders who spoke no English. To this twenty-year-old Swede from Seattle, they seemed exotic, handsome, proud. They were like refugees, people superior to the circum-stances in which they lived. I

was a young photographer hungering for a story I could tell with pictures, and the people of La Loma, Palo Verde, and Bishop allowed me to enter their lives with my camera.

At that same time, the United States was entering the McCarthy era, a period of bitter anticommunism when anything that could be viewed as social criticism was seen as unpatriotic and dangerous. In 1950, five of these photographs were shown at the Los Angeles County Museum in a national invitational show. Otherwise, there has been no audience for them until now. At the end of 1997, fifty of these images were exhibited at the Los Angeles

Public Library. Other shows have been scheduled, a film is in production, and the photographs will be published in the fall of 1999.

The year after these photographs were made, people began to be evicted from their homes. "To make room for low-cost housing," they were told. The housing was never built, but a few years later, Dodger Stadium occupied the land. People who lived in the neighborhoods have formed a group called *Los Desterrados* (The Uprooted). They still hold an annual picnic where the destroyed past is remembered and their continuing sense of community is celebrated.

# CHAPTER 4

**El movimiento**
# The Chicano Cultural Project Since the 1960s

Tomás Ybarra-Frausto

*in conversation with Michael Dear*

**MD:** How would you describe the Chicano cultural project since the early 1960s?

**TY-F:** I think the first thing to say is that this is a very personal reminiscence. The project was collective, and those were halcyon days in which many of us became conscious of creating ourselves as historical subjects. The Chicano movement of the sixties was a recuperation of a project that had been going on since 1848, with labor strikes and mobilizations and people fighting for justice, dignity, and human rights. The difference between previous social movements and the Chicano movement was one of scope. That is, it was a national mobilization of people in New Mexico, Colorado, Texas, New Jersey, and California. It was a collective. It was also like a mythical movement—it was as if this fire of your identity was slowly going out. Then there was this huge ignition, as if the fires were rekindled. It was cleansing at the same time as it was, for some people, destructive. So, the Chicano project of the sixties was the continuation of a long series of struggles of people in the United States. It was different because it was national in scope and also because it had this mythic quality about it.

For me, one striking thing was the involvement of creative people. Whereas before, there had been labor strikes with community activists, this time artists, intellectuals, musicians, and poets all joined in. This is what gave the sixties' movement its particular aura of transcendence and deep resonance. People activated ancient myths and symbols. It was a revitalization of the real symbolic capital that people had. This participation was absolutely different and was focused on everyday life practices. Our art was articulated at that moment. It came from the calendars we had on our kitchen walls, the *corridos* that uncle sang around the table, the stories that were passed on by grandfathers and grandmothers. The artists took this ordinary stuff that was part of our lived reality and elaborated on it through their own visions. Our art was founded on everyday life practices—the pain, the passion, and the beauty of life in the *barrios* and the urban spaces but also on the regional, old places where people still maintained many different types of traditions. For example, in New Mexico, one of the really important traditions was the whole *penitente* tradition, based on an outlawed church in New Mexico and the ritual songs of the *penitentes*.

This was a moment when everybody was traveling. I was living at that time in the Northwest, and yet I was going to California. We were all young. We were all idealistic. We all had a duffel bag and took the Greyhound bus and went to meetings throughout the country. That's how we made connections. Many of the artists of the Chicano movement (who are now recognized international artists) were part of this network of people crisscrossing the country, talking, never sleeping.

One of the important writers of the period was Tomás Rivera, who wrote a book that galvanized many of us early on. It was called *Y no se lo tragó la tierra*. I think it's been translated as *And the Earth Could Not Part*. It was a book about the farmworkers' experience, people traveling from Texas to Michigan to work in the fields—the racism they encountered, and their will to live. Tomás Rivera said that the basis for the Chicano creative act was three words: (1) *recuerdo*, remembrance; (2) *descubrimiento*, discovery; and (3) *voluntad*, volition. Remembrance, discovery, and volition. Although there are many ways to characterize the cultural project, these words serve very well.

The project was to recall (remember) what made us, who we were. The most ordinary vocabulary and words of our existence were brought into play, for example, the word *cariño* (which means love or gentleness or tenderness). We would have long discussions about what it meant and about the key words that are the ethos of community, such as *dignidad*, dignity. What does it mean? Who has dignity? Do buildings have dignity? Do spaces have dignity? How do you acquire dignity? Who gives you dignity?

It was a philosophical, but at the same time an ordinary, discussion of words that people use everyday. We started talking about their meaning. Then these words began giving shape and form to works of theater, or an artist would set out to write a story to explain what they meant. We would just talk about ourselves and remember, for example, how it was to get up early in the morning and hear your mother making tortillas in the kitchen. You were snug and warm in your bed and feeling that this was the most beautiful space that you had, with the tortillas being patted in the kitchen and the radio playing in Spanish. Then you would get up and were brought into the circle of love. All of us remembered the good, because the larger society was telling us that we were dirty Mexicans who didn't have culture or a history.

So, we inscribed our own meaning, remembering a past that was a basis for the meaning. It was a discovery of language and tradition, but more than anything, I want to emphasize the ordinariness. You remembered a bicultural reality—how you used to like peanut butter on *tortillas,* how when you took *chorizo* sandwiches to school, some of your classmates would exchange a *chorizo* for a sandwich. One tortilla was worth three sandwiches. So, you begin chuckling, remembering how you asserted yourself and made a space for yourself.

**MD:** Where were you growing up at this time?

**TY-F:** In San Antonio, Texas. I come from two distinct groups of people. My father and grandfather were ranchers; my mother's folks came from the north of Texas, and they were sheepherders. So, in my family, there were two traditions: the extroverted, open *vaqueros* with their own stories, *corridos,* and their style of boasting; and then the quiet introspective shepherds, who go out for six months and see no one and have their own songs about nature. So in myself, I have a rich bottomless pit of two traditions that have particular ways of looking at the world.

In the movement, we were all putting together our own worldviews. It was a wonderful mixture because some of us (like my family) were fifth-generation Mexican Americans who were in Texas while it was still part of Mexico. Other comrades had just come from Mexico. So the stories of the revolution, the stories of migration, the stories of San Antonio were all mixed in together. I had a very good friend, and we did everything together. But every year at a particular moment, they would close their house up, board up all the windows, and their garden would go to seed. I couldn't understand why he would abandon me, but my mother would say: "Well, they went up north *a los trabajos.*" They went up north to work. When Tomás Rivera's book came out, many of us understood what it was to go up north to work. We were urban, and we were settled, but many of the Chicano community were migratory. They would go wherever there was work, primarily agricultural work.

The first word is remembrance. The great rememberers were people such as Luis Valdez, who did the Teatro Campesino based on a mythic remembrance of farmworkers who had struggled. But it was the same workers of the Chichimecas, or the Aztecs struggling against the imperial powers. So, Valdez constructed a mythic story in which the gods were ordinary people. Even the political actions of, say, Cesar Chavez and the mobilization of the farmworkers, had a mythic base. The *peregrinajes,* the pilgrimages that Chavez originated, were long marches of people through an arid landscape that they would make fertile again. The farmworkers' eagle is a black thunderbird. The eagle in pre-Columbian mythology represented the sun, but the farmworkers' eagle was black, meaning that the people had to act, and through their actions they would bring back the sun that nurtured the crops, which gave us our food, which made us human. If you turn the farmworkers' thunderbird eagle upside down, it's a pyramid.

It was as if people were re-membering, that is, putting together the members of the community. They were re-membering community, yet they were forming a new body, a new consciousness, forming a new self that was the *Chicano self.* Remembrance was a big part of the Chicano movement, remembering the stories, re-membering the landscape. Landscape was fundamental, whether it was poetry about the *nopales,* the trees that you grew up with. Whatever landscape it was, people remembered because it was part of who they were. It was a landscape of the imagination, because although our deeds were real, we made them imaginary through their elaboration.

**MD:** In your remembering, did you look further back for roots, even to Spain?
Was this part of the tradition?

TY-F: Well, if you lived in San Antonio where you're surrounded by seven Spanish missions, you couldn't fail to recognize that reality. Of course, it was filtered through a Mexican sensibility! For some, Spain was anathema because Spain represented the Church, and Spain represented the land-owning class. On the other hand, Spain was represented by the very language that you spoke. Spain was represented in the Laws of the Indies that described how a settlement was to be begun. And you lived in that kind of settlement in your own *barrio,* the grid with the church, houses around it, and streets that radiated from it. So, yes, Spain was always present in the Southwest, and particularly in Texas. And yet, as you told the traditions, Spain became reintegrated with all kinds of other elements, a beginning of this *mestizaje* of multiply faceted things that were all in one place. This was part of the discovery, *descubrimiento,* that any tradition was a hybrid.

I grew up in cotton-picking country, and the women wore sun bonnets, *garsoleas.* The *garsoleas* from one part of Texas were shaped differently from other parts. Much later, in Spain to look at regional dress, I realized the basis for the *garsoleas* was the traditional headdress that women laborers in Spain wore. What was important wasn't the grand traditions of literature or music, it was the ordinary traditions of dressing oneself, of making an altar, of singing a song. In Texas and New Mexico, for example, some people sing the *corridos,* which are New World stories. But there are still some people who sing *gestas,* the long Spanish ballads that are about kings, or stories of who was the father of whom, who begat whom, and so on. The Teatro Campesino put together the medieval *pastorelas,* the nativity plays. It's a simple story of the shepherds going to see the Christ child, but these are ancient traditions that came through the Spanish Church to Mexico and the Southwest.

So, Spain was very much a part of our consciousness, but it was filtered through a Mexican and a Mexican American reality. In San Antonio, when young girls went to dancing school, they didn't learn Mexican folkloric dance, they learned Spanish classical dances with castanets. There were great teachers of Spanish classical dance during the fifties! That isn't very long ago, but the power of this cultural project was that in two decades, it imposed itself as a *mestizo,* a hybrid, a negotiable reality where you can take from Spanish, indigenous Native American cultures, and all the other cultures, and then weave your own. This was the discovery; this was the great *descubrimiento:* that in every region, in multiple ways, people could take different elements to produce something new. This freed the imagination to borrow, to create.

Of course, the Anglo American reality was part of this process, too. We grew up going to Anglo American schools learning Anglo American poetry. I loved Emily Dickinson so much. Later, when I came to the East Coast, people couldn't understand why I took a bouquet of flowers to her grave, but Emily Dickinson was the first poet that I had learned. I knew the Spanish *corridos,* but I learned to read poetry first in English. And I read Emily Dickinson. Emily Dickinson is as much a part as Gabriela Mistral became a part of me. Discovery was like two tectonic plates coming together. It was a cataclysmic, volcanic eruption of recognition of who you were. You who had been accused of having very little, actually had too much. You were told lies that you were ahistorical and that you had no culture, when the reality was that you were immersed in a deep sea of cultural histories, and you could make meaning for yourself.

When you put together the deep meaning of all these ordinary things, you begin realizing that the way beans tasted, or the way celebrations happened, were events that created structures of social meaning, structures of family, and structures of opposition and resistance. Such structures gave you the power to resist the dominant culture. Language was especially important. In Texas, we were always bilingual. We spoke Spanish at home, but we heard radio (television was just coming in) in English. I was used to going to the theater and seeing Esther Williams in the morning and María Félix in the afternoon. We were totally bilingual, but the idea was that you couldn't have a divided loyalty. There was room for only one loyalty, which was to the English language. So, in my elementary school, there was a language patrol. It was a big honor because you got to wear a shield, and during recess you would take the names of your classmates who spoke Spanish and report them to your teacher. Their names went up on a roster of the people who committed the crime of speaking Spanish. This made a big impact on me because before I went to school, my father had shown me a map of Latin America. He said: "Let's look at how many places speak Spanish." We started in Chile, then came to Venezuela, Mexico, Central America, and San Antonio. "Now, let's look for the places that speak English." And we looked at Canada and the United States, and a little English-speaking colony in Central America. It was quite clear that the largest group of nations spoke Spanish, and my father said: "That's why you have to speak Spanish, because you're an American. If you're American, you speak Spanish." So when I went to school the first day, the teacher said: "Well, we're all Americans, so we all have to learn how to speak English." I raised my hand, and said: "Teacher, no. We're Americans and we have to speak Spanish!" That was how I ended up with my head down, and in the corner because I was insubordinate.

Taken together, small insubordinations became the catalyst for a social movement. The Chicano cultural project was a fight against an attempt to delimit you and make your space small, when in reality your being and meaning were rich and large. This was the way that remembrance led to volition. Volition (*voluntad*) was the will to act, but it was also the recognition of a long historical trajectory. I'm speaking personally, and this was my story of how I learned to read. I had an aunt who was a *magonista*. The

Magon brothers at the turn of the century were incendiary figures in the Southwest who believed in the liberation of Mexico, poor people, and the land. They mobilized many Chicano communities, and they had a newspaper that I used to learn to read, letter by letter, before I went to school. I was five years old, and my aunt would articulate and I learned. My mother would laugh later on when I was very involved in the Chicano cultural project. She said: "Well, the circle has been closed because when you were a child, the first words you learned were *huelga* and *pueblo* and *comunidad.* You would articulate *co-mu-ni-dad,* and it was from the Flores Magon newspaper." So, there was a long tradition of cultural and political activism against injustice going back to the Mexican revolution. Again, the words were very simple: *pan, justicia,* and *libertad*—bread, justice, and liberty. These words became articulated into the Chicano movement as a continuation of the struggle for the basic rights.

That was the project: remembrance, discovery, and volition. We haven't completed the project; we have not yet finished discovering. It wasn't the great theoreticians in mural painting, and it wasn't the great writers of the Mexican revolution. It was all of them, but more important, it was how our parents, grandparents, and families had articulated the meaning of bread and justice and liberty. I think that as the cohort of the cultural project of the sixties, entering the *tercera edad,* the third age (meaning we're getting older), we're still struggling because we have yet to create an aesthetic out of all these wellsprings of knowledge, that in many ways are still *subterrados.* In learning how to do things, you maintain them, but every time you touch them, you change them, because you add elements from your own reality. In other words, we're learning how alive and deep all these hybridizations and traveling cultures are. And we're expanding the repertoire not only to Spain but also to the Mediterranean and beyond. We're now at the point where we're articulating how these major cultural knowledge systems influenced our architecture, planning, building, living, and dreaming. It's hard because I'm talking of a bottom-up kind of history, art, architecture. It's a beginning, but there is a lot to be done.

**MD:** Do you think that the cohort of the nineties sees the unfinished cultural project in the same way as you do, or is there a dichotomy between the new and the old?

**TY-F:** I'm very excited about the cohort of the nineties. I think that they've learned their lessons well and that they're going much further than we did. Our cohort in the *Movimiento* of the sixties is also critiquing the project. It's a gentle critique because our very being was invested in the project. We were monolithic in thinking about culture. We tried to amalgamate everyone under certain types of strictures. We tried to think that New Mexicans and

Californians and Texans in some ways were struggling for the same things. In one way they were, but in other ways they weren't. We had to create this imagined community that was whole and holistic, and total and energized. We *had* to do this to struggle against a dominant culture that made you segmented and unaffiliated. The notion that we were united was, in a way, a myth. Now the younger generation asserts that it's all right to have fissures and breaks and contestation about identity, meaning, and worth. Having this contestation does not mean that you are disunited. You are simply united in a different way. But we tried to think of culture in a holistic, totalizing way. It was a mistake because we didn't allow for the particularities of experience to come forth— a particular way of speaking, dancing, or singing that varied from one region to another. We made it all a *Chicano monolithic.*

**MD:** But wasn't that necessary at that time?

TY-F: I think it was. But these types of monolithic ways of seeing can lead to narrow strictures and racisms based on particularities: "I'm like this, and if you're not like this, you don't belong." We were open, but we were not open enough. Today's generation, the people who are articulating the project of the nineties, understand that things are negotiable. For example, we said we're *Chicanos,* which means we're not Mexicans from Mexico and we're not Anglo Americans, but we're a mixture of both those things. Today's *Latinos* are saying that we are all the places we come from, plus the experience of living in the United States. We're negotiating and synthesizing all these experiences. This opens up all types of possibilities in the cultural sphere because you have such a wealth and such a multiplicity.

The question is: How can you have so much difference *and* a common ground? What do you have to give up to be a part of a common project? As I said, we were too monolithic in our conception of culture, instead of seeing it as open, permeable. Being Chicano is something constructed, and this is what the new movement is grasping. You can change your identity. One area in which the Chicano movement was very closed was sexual preference. The gay and lesbian communities were participatory in the movement, but silenced. But now these voices are creating their own vision and their own reality in the new project.

Another critique came from that many of us who made up the cultural projects of the sixties were working class. We were very proud that we were workers, whether in the fields, factories, or sweatshops. We also had people from the middle class and elites, but I think that we romanticized the working class. Now, slowly, we are recognizing that many of our friends who came from the middle class were very much part of the Chicano movement. But back then, they had to play a game—that they came from the working class—to belong to this massive mobilization. A recognition being made by the new cohort is that the middle class and the elites can contribute if you can mobilize them. If you can find those things that all classes can fight for, this makes for a stronger struggle. Many of the cohort of the sixties are now quite well ensconced in the middle class. What matters is how you *use* your privileges for the good of the most fragile and most needy.

A last critique has to do with nationalism. The Chicano movement based its ideology in the ideals of the Mexican revolution, the Cuban Revolution, and Third World struggles. We were very conscious of what was going on in Africa, Vietnam, and South Africa. But essentially, ours was a nationalist project. The new project of the nineties is a global project. People are seeing and making connections with Latinos in Latin America and Europe, asking how exiled communities belong in the continental project, the national project, or the local project. The project for the nineties no longer is localized in a nation but is a transnational global project. And how you can be global yet still local is the main responsibility of artists, city planners, designers, dramatists, and intellectuals as they articulate this new project.

**MD:** Has this globalization process weakened the links between the movement of the sixties and the movement of the nineties?

**TY-F:** No, I don't think so. As we begin articulating the variations within the Chicano movement (*Norte Mexicanos, Californios, Texanos*), we discovered such a wealth of cultural practices. The design of a garden in each region was similar but yet distinctive. The *local* ways of doing things were very significant, and paradoxically, these local ways are going to be very necessary because globality tends to homogenize. That was something we were struggling against in the sixties as well: We were homogenized as Mexican or American, so we became Mexican American, which gives you two options of being. In the nineties, people are arguing about multiple options. The local is going to be much more significant. So, many lessons from the sixties are going be useful to students and activists of the nineties.

**MD:** But the contemporary discourse must be different, because the terms of reference are so much broader? The old canons are being diluted, maybe even rendered obsolete?

**TY-F:** It's always the privilege of the old to think that the young have lost something. So, taking that privilege, I think that one of things that has been lost is the depth of connection with the points of origin. Of course, the ultimate point of origin is Spain. Whatever ideological impediments that invoke it, it is still there. The patterns that led to the encounter with the New World reverberate to this day. In the sixties, we took courses in Spanish culture because Latin American culture was in its infancy. Now, of course, there are Chicano and Latino studies. Very few are opting to take Spanish culture courses. But it's there that we learn about the Great Generation of '98, what it meant to poets and philosophers in Spain to lose their dominions in the New World. So we were conscious of the vanquished as well as the victors. Learning what it meant for those who lost humanized the project of the sixties. It wasn't that Chicanos wanted to be victims, were victims, or wanted to be victors. But we needed to know the victims, their suffering, and the meaning of victimhood.

This is one way in which the dialogue is getting smaller, the recognition that those who lose are as significant as the winners. So, what did it mean when Mexico lost one third of its territory and one tenth of its population after the end of the Mexican-American War? Well, until recently, only the victors had written about that war. But Chicanos were the *results* of that transfer. I'm wondering whether the current generation understands that. The conversation *is* getting thinner in major traditions, yet we belong to the Western tradition. We have changed it, reinforced it, battled it, but through the great traditions of Spain and Latin America, we are part of that tradition. But people are paying less attention to it. That is what I mean when I say that the conversation is getting "thinner."

**MD:** So, where do you see the Chicano cultural project heading in the new century?

**TY-F:** Well, the project is a new project. It is a Latino project. Yet I think we have a lot to learn from the uncompleted Chicano project. The most basic thing is still to reach deeply into the ordinary lives and experiences of people and to extrapolate from those lived realities to the creation of new forms of being and art. If we're going to have at least the possibility of a counterspace or a new public sphere, we must go back to the primordial questions: Who are we? What do we want? How are we going to achieve it? We gave partial answers to these questions. For example, one of the things that the earlier project did was to reinvent symbols from Mexico. For example, the Day of the Dead celebrations were not part of U.S. culture until the artists of the Chicano movement started reinventing. We invented a tradition that had a basis in real Mexican traditions but was transformed in the United States. The Day of the Dead in the United States in many ways celebrates the real elements that are causing death and destruction in Latino/Chicano communities (e.g., AIDS and pesticides), whereas in Mexico, the Day of the Dead is a metaphoric, artistic revelation. So we're celebrating an ancient tradition of living and dying, but we're also paying attention to what is causing death and struggling against it. The figure that I would use to describe the nineties' project would be Janus—the old Roman god who stood on the threshold facing backward and forward simultaneously. The millennial threshold is of the past and of the future—the space between one movement and another. We have to look back: What was the unfinished business? On the other hand, we are looking forward to the virtual world of cyborgs! I'm optimistic that the newer activists can look forward in a freer, more negotiable world in which things are constantly being contested, re-formed, and rearticulated. Yet I hope they will also look back to stake out their positions after deep thinking. It's an open project, one that's ongoing.

**MD:** Your emphasis on the local stresses how important *place* is to what people do. Do you have any sense of the difference that place makes in the new Chicano culture?

**TY-F:** One of things that we learned from the sixties was that when you're displaced, you carry the place with you. I remember talking to many Chicano students at Yale, Princeton, Harvard—people who came from New Mexico, San Antonio, or Los Angeles. They would re-create a space with the music, food, and companionship of their village wherever they were. I think you recognize that the place is within you, and that recognition is very important. Place is irrevocably a core reality of identity, culture, and meaning. It's not only external, it's internal. It's not only geographic, it's spiritual. It's not only there, but here. You negotiate through all these all the time, as you enter new spaces.

**MD:** The reason this question is important is because you've expressed the contemporary project very much as a global enterprise. Yet under globalization, those place-based ties you've articulated must be getting looser. What does this mean for the future of the movement?

**TY-F:** In the sixties, remembrance had a way of being selective. You remember the wonderful things that happened in a particular place; you remember that the trees were three times bigger than they really are. It is important to articulate the bigness, the glory, the magnificence, the stupendous quality of those remembered places. Then you go back and find that the place is dusty and the trees are scraggly; you realize the magnificence is very ordinary. Even today, as you are flying, driving, or walking across borders, somehow your mind is avidly reconstructing and remembering the spaces you left as glorious places. You can't escape that. Space is fundamental to identity and being, and it's the "glocal" (the combination of local and global) that is the ground for new ways of looking at things. Where you don't privilege one or the other—that's the new reality. How the global and the local come together will define the emergent art forms.

There's also the paradoxical condition that as things open up, many people in gated communities are closing into themselves, hiring dogs and private security forces to maintain the space that is not part of the global. There are many other people, including elderly people, whose spaces are confined to their bed, dresser, and one window. In the social realm, many groups of people are disarticulating themselves from the public sphere, creating guarded inner spheres, afraid of the things that globalization brings—disease, communication, and interaction. The struggle in the cities, for example, may involve gated communities that have the means to isolate themselves versus those calling for open gates and a more public sphere.

**MD:** One criticism that the younger generation makes of the sixties cohort is that their vision was excessively utopian. How do you deal with this type of criticism?

TY-F: For me, the word *utopia* is a very good word. It's a positive word that is part of an imaginative vision, or a dream, the possibility of remaking yourself. Yes, the movement was utopian. It was also historical. The New World was a utopia compared with the old. It was conceived as a utopia by all those people who wanted to create (from multiple cultures and spaces) a unified sense of being. If we were being utopian, thinking we could unite the Mexican and the American communities to make a Mexican American hybrid, just think of the Latino project! They are trying to amalgamate twenty or more countries under a single rubric. I think that's fairly utopian. But . . . it's a continuing project. Utopias, for me, are the necessary realms of imagination, of spirit, of dreams.

New York City
March 1998

# CHAPTER 5

# Bars and Belonging

María Elena Fernández *(text)*
Jessica Chornesky *(photographs)*

On a Saturday morning in early December, I drive up a street of my old East Hollywood neighborhood, my hatchback packed with the first load to move from my Highland Park house into my new home in this Armenian and Latin American immigrant neighborhood. I catch a glimpse of the apartment building I grew up in a block away. It looms over a one-story commercial building, now a suspicious-looking dental office that looks as if buckets of stark white paint have been dumped on it, decorated with black graffiti.

What the hell am I doing? The six-foot-high black iron bars that have appeared around so many of the houses and apartment buildings since I moved away in 1987 suddenly grow taller, thicker, and blacker and begin to close in on the sides of my car.

The fears that I believe belong to my parents and friends now ricochet in my brain. "Why do you want to move back there?" "You mean you're moving from a bad neighborhood to a worse neighborhood?" "Isn't it dangerous?"

I always give the same answer. "Because it's what I want. I've always wanted to move back, and I'm finally going to do it." I am afraid they will laugh if they hear my real reasons: Because the neighborhood between Santa Monica and Western, Normandie and Sunset, is the only place in L.A. I feel I belong. Because this neighborhood taught me what community means. Because I want to be back where people speak different languages and even Spanish comes in different accents. Because I am enamored with the immigrant dream of adults sacrificing so that their children can have a better life.

But there's a secret that I won't whisper to anyone, scarcely myself: I want to come back and save the neighborhood. I want to save it from the ills I remember from my childhood, like the Pussycat Theater that I passed every day when my father drove me up Western Avenue to my Catholic high school. The morning I saw that the Chris and Pitts restaurant on Sunset had been turned into a topless club, my brain screamed, "Don't you know that families live here!" and I envisioned a neighborhood campaign to shut it down. And I had a frequent fantasy that late at night I would paint over the first two letters of the bold, black "Female Mudwrestling" sign that ran across the front of the Hollywood Tropicana.

As I pay the deposit on my new apartment and pack boxes, I dream about moving back home, the fantasies of my adolescence dancing in my head. But as I drive up to my building with my first load, the six-foot black iron bars overpowering the sidewalks disturb me. Some are disguised in green or white paint, with scalloped designs and decorative spear ends. But they all look like iron fortresses, prison bars built to barricade in the fearful. They are my neighbors' solution to the social chaos that descended here in 1992. A new fantasy springs up before my eyes: On a Saturday morning, hordes of my neighbors line the block, laughing and talking and together pulling the fences up out of the ground, declaring that they don't need them anymore, it's just that they got scared after the Riots. I decide I want to start a neighborhood association.

■

I grew up in East Hollywood in a typically unattractive 1950s matchbox building, before there were iron bars. I was born a mile away and lived there from the time I was six months until I went to college at age eighteen, sharing a two-bedroom, one-bathroom apartment with my parents, brother, and sister. I returned there for a few years after I graduated from college.

I was supposed to grow up in Mexico City. I was six, and after eight years of saving money to invest in my father's silver business, my parents were on the verge of moving back. But it was 1971, and silver prices skyrocketed, the business closed, and our landlord, Mr. Kaufman, had just offered to sell to the building managers: my parents. They shifted their savings to the down payment and anchored our fate to the United States and to the big white building with the olive trim on Harvard Boulevard. My father painted a metal sign orange with pine green script that read "Villa Victoria," after my grandmother, and hung it in front of the apartments under a tree.

The two-story courtyard building buzzed with the voices of my Egyptian, Armenian, Colombian, Cuban, and Venezuelan neighbors. Mr. Assadourian, our elderly Armenian neighbor, lived in the building for as long as I can remember. He was tall and sturdy, slightly overweight. He combed his longish gray hair back against his scalp and always seemed to have two days' worth of white beard. He almost always wore a yellowed white T-shirt tucked into dark and worn baggy dress pants. He was everybody's keeper—telling the kids to shut up when they got too loud, and, after I learned to drive, he'd peer down at me from his window every time I pulled into the alley parking lot late at night.

Mr. Assadourian was also an integral part of my family's food supply. He regularly went fishing before dawn at Redondo Beach to catch bonitas. He always gave my mom several—gutted and cleaned—that we'd have for dinner that day. He would alert my mother as to the bargains in the neighborhood markets and would call our house to see what my mother wanted before going downtown to Grand Central Market.

"Ask your mother if she want cantaloopie."

"If she wants what?"

"Cantaloopie." he repeated in his gruff voice. "Ask your mother if she want cantaloopie."

"Hold on . . . Mami! Te habla el Sr. Assadourian. Que si quieres no sé qué. No le entiendo."

I could never understand most of what Mr. Assadourian said. But somehow my mother always did. I could hear them talking and see them gesturing from our living room window when she visited their apartment across the courtyard from us. Somehow my mother returned with a report about how his son's jewelry business was faltering, that his grandson in Fresno was now playing baseball for the minor leagues, and that he was sure that the next-door neighbor's husband was an alcoholic.

Another frequent visitor to our home was Mrs. Tirado, our Ecuadoran neighbor from down the street. She'd come by two or three times a week around dinnertime to visit my mom. They would gossip in our small yellow-tiled kitchen while my mother prepared dinner. I loved it when she brought us mondongo, the spiced tripe dish from her homeland, or papa a la Guancaina, potato smothered in a tangy golden sauce from Peru, her husband's birthplace.

In the early dawn hours, they would meet again, my mother driving by to pick up Mrs. Tirado as they went to their jobs together as janitors at Disney's Burbank studios. Mrs. Tirado got my mother that job in 1978 when my mom had to work for the first time because my father fell ill and could no longer work as a silversmith.

■

I had combed the neighborhood and found a vacant apartment in one of the few buildings with some semblance of beauty and

no iron bars. The muted rose structure badly needs a paint job, but it is beautiful—a Spanish-style apartment building with arches lining the courtyard and hand-painted tiles on the red cement steps leading up to my second-floor apartment. Inside, the ceilings are high, and the rooms feel vast. Poinsettias reach up to my breakfast nook window, and from my living room, I can see the persimmons in my neighbor's backyard, a fiery ripe orange, starting to fall from the tree. Every time I drive home, I make each turn without thinking, and like a homing pigeon, I follow my route with complete familiarity. Despite the noise of the Hollywood freeway like a distant rushing river seeping through the window, I feel serenity here. Finally, I am home.

The courtyard in my new building is almost always empty when I come home from work or throw out the trash. I hardly see my neighbors. Ogegeko, my African American musician neighbor with the dreads halfway down his back, is the only one I see regularly. The son of my Guatemalan neighbors, Zoila and Exsau, isn't out much. Occasionally, I bump into my Salvadoran neighbor, Marina, who takes care of the stray gold cats. A few times, I see Ogegeko's partner, Anita, and she greets me warmly. Very rarely do I see Sybil, the older Jewish woman who's lived here for twenty years, or the four young Zapotec men who live upstairs. Four of the twelve units are vacant. Only swift greetings are exchanged, let alone plates of food across my doorway. I remember that when I went to look at an apartment building around the corner from where I now live, an elderly Armenian man standing outside said to me, "Good building. You'll like it. There are no Mexicans here." And there are no Armenians in mine.

I shove my disappointment out of view and instead concentrate on my mission to organize a neighborhood association. But I scale down my plan to a tenants' association for my building. Two weeks after moving in, I start talking to my neighbors about having a Christmas party, and the planning meeting is scheduled for a Saturday morning at my place. Lucina, Ogegeko, Zoila, Paquita, and the manager all come.

"I'll make taquitos."

"I'll buy the drinks."

"I can bring a cake."

"Do we want to get a piñata for the kids?" I translate as needed.

On the day of the party, the courtyard is overflowing with food, the manager surprises us and brings potted poinsettias to decorate the tables, and the kids gleefully demolish the piñata.

After the party, everyone retreats back to the apartments, and the courtyard is empty again. I see the Liras, my neighbors on the same landing, more frequently than my other neighbors. They're a couple from Mexico City and live with their teenaged kids and granddaughter. I try to get to know them by asking them for favors. Do they have a hammer I can borrow? Can Moisés and their son Ulises help carry my new couch into my apartment? Does Lucina have any advice for me on getting rid of the roaches I've found? She had them, too, and the owners wouldn't fumigate. Try boric acid, she suggests. After she laid it down, they never came back. I go to Sav-On to buy some. Four teenaged brothers and sisters are in the same aisle picking out a box of rat poison.

One afternoon, I bump into Moisés at the top of the stairs. As I'm about to walk into my apartment, he stops me, "Listen," looking down on the ground, "I wanted to ask you if you know of any jobs, maybe where you work. I just got laid off." I don't know of anything, but I promise to ask around. I remember the kids at Sav-On and the clans of Jehovah Witnesses who come to my door almost every weekend. Last time, they left me a booklet titled "When No One Is Poor."

This is not how I remember it. Growing up in Villa Victoria, my neighbors seemed to have stable white-collar jobs. I remember Edgar Sánchez dashing out of his apartment, looking dapper in a suit to sell insurance. Sandy was a secretary, and Fanny, our gregarious neighbor who wore a miniskirt anytime she could get away with it, worked at a bank. Until he got sick, my father earned twice the minimum wage as a silversmith, which paid for our Catholic schools and for my mother to stay home and take care of us.

■

In April, there's a rash of graffiti at almost every corner I turn: the tags of the MS13, the Salvadoran Mara Salvatrucha gang. One day on a late afternoon walk, I turn the corner of Normandie and Santa Monica and see an Asian youth crossing out a rival's name on the mini-mall wall facing the street. After he jumps into a car with three other young men and drives off, I approach the wall to look at what he wrote. As I leave, I see that they have stopped their car in the street and are looking menacingly at me. I walk up Normandie, nervously checking each approaching car. I am afraid they will pull up and confront me.

Two blocks away, Mara members are hanging out at the phone booth on Fountain across the street from a liquor store that has become a center of the local drug trade. My mom used to send me to this liquor store on Saturday mornings to buy milk so she could make pancakes. I am afraid they will sense my fear. I try to call up my usual compassion but instead wonder if each of them is packing a piece.

Another day that week, I am walking on Santa Monica Boulevard to the commercial bustle on Western Avenue. I pass a tall, disheveled white man who looks harmless, but definitely high. Then I walk by a man I've frequently sighted from my car. He wears glasses; a baseball cap over blonde, curly shoulder-length hair; and socks and tennis shoes. Hormone-induced breasts stick out from under a cotton blouse. His legs looked like rubber, an unnatural tan color with varicose veins protruding from a short, plaid, pleated skirt.

Still jarred from seeing him, I suddenly notice a little boy about eight walking next to me. He has light brown skin and thick curly hair and wears a white karate outfit. He's carrying a navy tote bag on his shoulder, about as long as he is tall. I look around for an adult who would be with him. There's no one.

"Are you by yourself?" He nods. "Where's your mom?" He gives me a long explanation I don't entirely understand, something about her selling things. He is going to Lemon Grove Park to meet her. He explains that he is coming from his karate class "over there," turning around and pointing. I try to recover from my shock.

"Well, be really careful, OK? I'll walk you part of the way, OK?" He nods. "What's your name?"

"Martín."

"I'm María Elena." He smiles up at me. "How old are you?"

"Seven."

When we get to a small street, he tells me this is where he needs to cross. The 6 P.M. Santa Monica Boulevard traffic is bolting by in both directions. There is no traffic signal.

"It's better if we cross back at the light." I wonder if he ever tries to cross at this intersection or if I just distracted him. We walk back a block to cross at a signal. As he turns down the residential street toward the park, we say good-bye. I turn around again to watch him, and he is looking at me as he walks down the street. I wave again, and he flashes a huge smile and a long enthusiastic wave, then grips his tote bag and trots down the block.

My God, if Martín would talk to me and be so trusting, he would talk to anyone—any one of the strange people I had passed. What will happen next time? What will happen to him when he grows up? Will he be working the liquor store parking lot? I go home that night and can't stop thinking about Martín. Depression shoots through my body and leaves me languishing on the couch for the rest of the night.

I start to wish that I hadn't moved back; my warm memories are getting washed away by a 1990s reality seen through my adult eyes. I start to wonder if I belong here. Why did I think I could "save" my neighborhood? Maybe I should have moved into Los Feliz with the other yuppies and aspiring artists. Is that what I really am? All I know is that as I walk up my street, I see only black iron bars lining the sidewalk—even when I close my eyes.

■

After a period of mourning, less idyllic memories fill in my childhood landscape where the Pussycat Theater and Hollywood Tropicana loomed. Patty was my best friend and next-door neighbor when I was in the fourth and fifth grades. She lived with her

mom. She sometimes dressed in chola fashions popular in her public school: light blue corduroys and a black elastic belt with a square silver chrome buckle with a capital *P* for her last name. The cords went with black hush puppies, hair parted down the middle, and simple twists of black plastic bracelets. I wished my mom would let me dress like Patty and walk home from school by myself like her.

Patty taught me how to do the 18th Street sign by crisscrossing my forearms. We'd watch the TV cholo dance program "The Huggie Boy Show" together on weekday afternoons. She was just a wannabe, but those were Clanton and XVII Street days, and she was afraid. "What do I say if someone asks me where I'm from?" After the sixth grade, her mother moved them to Riverside County.

I was afraid of only one person in my neighborhood: Ricky. He was the young drug-addicted brother of my neighbor Sandy, who lived with her son. I'd run into him when my mom sent me to the liquor store on Fountain when we ran out of milk or behind the building at night when I went to get the clothes out of the dryer in the laundry room. After a while, Molly kicked him out, and as far as we knew, he was living on the street. Ricky was dark skinned and skinny. He sauntered when he walked and always had a red rim around his eyes. He greeted me confidently, looking me in the eye, smirking, seeming to revel in my fear. I always avoided his gaze, barely said hello, and moved as quickly as I could.

I was too young to understand what rape was, but I thought if anybody raped me, it would be Ricky. After a few months of moving in, driving down Fountain near the liquor store, I see him on the sidewalk. I turn on the first street, afraid he'll peer into my car and recognize me and know that I have moved back.

■

I have lived here three years now, and reality has finally settled in: I live in an ugly neighborhood, the kind of neighborhood most people want to stay the hell away from. Before moving back, I reveled in nostalgia and cocooned in denial, seeing only the idealized community of my childhood. But it remains the most salient memory of growing up because I was protected from the neighborhood's dangers. I was swaddled in the affirmation given by my parents, the Catholic schools I went to, and my neighbors at Villa Victoria. They all pushed me out into the world prepared enough and strong enough to get two college degrees. Saving the neighborhood was the least I could do.

But I have failed miserably. For the first two years, I obsessed about it: talked to tenant organizations, requested materials, talked to my neighbors about problems in the building. The only meetings and projects that followed took place in my imagination. And I berated myself for not doing anything except organize two Christmas parties.

But I can't. My struggles are already too much for me, I finally have to admit. I am paying my bills from a part-time day job. At the same time, I am inundated in projects and deadlines as I try to forge a career as a writer. And still I'm always trying to find time for family, friends, and some rest. Last Christmas, I was so broke and so depressed about it that I didn't do anything about the Christmas party. How fragile I feel; how human, not heroine.

I try to tell myself that it's enough that I live here, that I'm a good neighbor. I'm grateful for the warm greetings in the courtyard, knocks on the door when my car's parked on the wrong side of the street, when a college catalog arrives for the Liras' son. One time, Lucina came over and wanted four of us to get together and ask for a rent reduction. She scheduled the meeting, I wrote the letter. We got the reduction.

That sense of community from my childhood has never re-emerged, and I am still nostalgic for it. But I continue to live here because as I ask myself the daily question of who I am while the outside world constantly tries to pull me into the whirlwind of buy more-own more-be more, this neighborhood connects me to the deepest parts of myself. In my neighbors, I see the reflection of the immigrant struggles that shaped my beginnings. And I continue to live here because it remains the best context for my own struggles.

# CHAPTER 6

# Tres poemas

Gloria Enedina Alvarez

**Vende futuro**

Cruzamos "la línea" entre las cuatro direcciones,
puntos comunes en caminatas largas.
We cross "the line" between four directions.
common points on long paths,
those same steps in other places.
los mismos pasos en otros lugares.
vivir, amar tan duro, tan abierto.
to live, to love so hard, so open.
Breathless Heart, Corazón Sin Suspiros.
pasos tomados por caminantes distintos
Ayer como Hoy
steps taken by different travelers
Today like Yesterday.
I saw him selling oranges.
Was that you, Earth Star?
Lo vi vendiendo naranjas.
¿Fuiste tú, Astro Terrenal?
Defying the cops with your presence.
outslicking them while you bury the fruit in the bushes—
The Invisible Man
El Invisible—
desafías a la policia con tu presencia,
ganándole, enterrando la fruta en las matas.
I saw him handcuffed the Caminante.
el sol amaneciendo en sus ojos cafés,
matizando la sonrisa,

AUTHOR'S NOTE: Copyright © by Gloria Enedina Alvarez. "Vende futuro" and "Contrastes//Contrasts" were previously published in *Chicana Creativity and Criticism: New Frontiers in American Literature,* ed. María Herrera-Sobek and Helena María Viramontes (Albuquerque: University of New Mexico Press, 1996).

muy esperanzado hace un día,
consiguiendo sólo para otra caja de naranjas a cinco dólares
confiscadas hoy como evidencia
Making only enough for another five dollar box of oranges
today, confiscated as evidence.
Invisible Criminal no name, no data, no land
Criminal Invisible sin nombre, sin datos, sin tierra
Smog naranja gris nubla sus dientes brillantes
hoy desaparecido, empacado
desde la Casa de Vidrio al Centro de Detención con la Migra
pérdida de mayoreo
orange grey smog clouds his shiny teeth,
disappeared today, packed away,
from the Glass House to the INS Detention Center,
wholesale loss
naranjeros color a sol, perfume cítrico ahumado,
naranjeras preñadas con un niño en espera en el carrito de market.
a un lado, en la salida del freeway, tu salida, Sr. Cultivador
orange sellers, sun colored, smoked citrus perfumed.
pregnant orange sellers, a child waits in the market cart
by the side, at the freeway exit, your exit, Mr. Grower.

■

Contrastes/Contrasts    Interminables, interminable silver gray cylinders
reflecting their cool glitter
against aging brown and brick red porous rectangles
now dwarfed and anchored on skid row.
Its slick disinfected shadows slip over
the wide matte-finish corners of the collapsed tent city,
as if masking los olores de vida, the smell of life,
perfumed with pungent mustiness of yesterday,
of living for the moment or momentarily living
with horns, screeches, bottles breaking in harmony
with the scattered motion of trafficking time, sex zombies,
living hallucinations manufactured in the dream factory
of the money gods,
children's cries winding up hotel staircases
to numb the heavy whispers of its needled residents,
pained souls begging food from the great city's garbage dump,
robbed of their essence.
The invading glass giants thought their hearts lie buried
under Bunker Hill, while they swept aside part of Varrio Diamond
to be replaced by its daily visitors in confusing tones
of black, white and gray.

They come casting their shadows but always
taking them home at dayend.
The Diamond Curse—Brillantes Vidriales—freezes the hearts
of its inhabitants, it's said.

■

Mujer del lago

Por el espejo, por el cristal
Del Lago de Plata
surge un hilo de luz
la esperanza de su rostro
es la ternura del abrazo humedo

Woman of the lake
upon the moonlit sheath
shadowed pine and eucalyptus
with smiles of anguished fish
whose eyes ignite
ripples of tinted fear
that rise from her tears
and lift the light
from her inverted sky

From the Lake of Silver
surges a thread of light
tenderness the wet embrace
in the face of hope

Mujer del Lago
con vestido de luna en manto seda
a su sombra pinos y eucaliptos
sonrisas de peces
con ojos que incendian
ondas en agua con tintes de miedo
alzan sus lágrimas
levantan la luz de su invertido cielo

# CHAPTER 1

# La Movida

Pedro Meyer *(photograph)*
Rubén Martínez *(text)*

Yep, Indians alive!
Not frozen in anthropological diorama
but servin' nouvelle Mexican on Pico Boulevard
and tattooin' whitegirl bellies on Melrose!
Dreamin' in Michoacán they owl-fly red-eye to L.A.
to rid migrant-daughter of *el mal de ojo*
and back again wakin' curled up in satellite dish
as dawn breaks in land of the butterfly!

And Cowboys they dyin' to hold the line & keep Injuns back
but there Meskins go re-discoverin' land
soaked with the blood of memory
re-stokin' fire of extinguished American Dream!

Rewritin' headlines, upsettin' wage-scales,
tonguin' English into a historical loop
they come endlessly becoming
breathlessly arriving
from Zamora, Michoacán
to North Hollywood, Michoacán

And now they be wearin' Cowboy hats
saunterin' 'long Broadway . . .
so if They've become Us, then Who are We?

SOURCE: The image has previously appeared in *Los Angeles Magazine* (1997) and in the Los Angeles Metro Transportation Authority (MTA) Bookmark series (1997).

# CHAPTER 8

# Manuela S-t-i-t-c-h-e-d

Christina Fernandez

DBA and TT&T Fashions, 237 and 239 San Fernando Road

Looking down,

Leo Fashion, 3011 North Main Street

she saw that her stocking had a run. "La migra came like a

3452 & 3450 North Pasadena Avenue

storm today." The end of a black thread was caught on her

San Pedro Fashion, 1409 North Main Street

heel. It trailed away, winding around the corner. She pictured

Napa Apparel Inc., 176 San Fernando Road

an empty spool and feared they would notice it and find her.

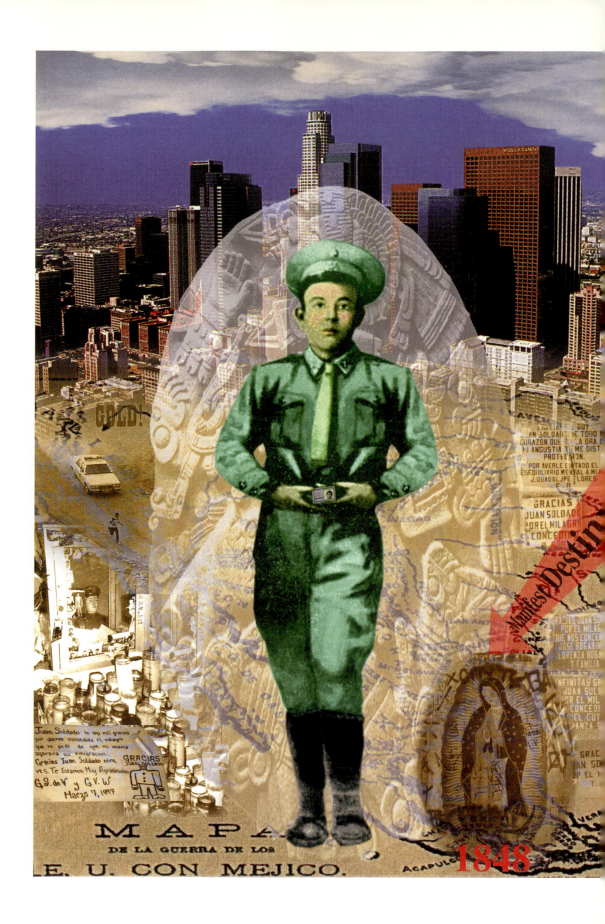

# 9 Juan Soldado

Alma López

Juan Castillo Morales, most commonly known as *Juan Soldado,* is the unofficial patron saint and protector of undocumented immigrants. Morales was a soldier in Tijuana accused of rape and murder and executed in 1938. He claimed that he was framed by a superior officer who actually committed the crime. According to the legend, he swore that he was innocent and his innocence would be proven when miracles were asked and granted in his name.

Today, people travel to Panteon 2 in Tijuana to ask for miracles relating to issues of immigration: crossing the border safely, dealing with the border patrol, and negotiating permanent residency and citizenship. Those same people are the ones who have converted his plain tombstone into a one-room altar space filled with photographs, letters, gifts, constantly lit candles, fresh flowers, and Xerox copies of micas (green cards). He is not recognized by the church and is therefore an illegal saint of "illegal" immigrants. Juan Soldado is about the creation of myth, spirituality, and cultural history as a survival mechanism during transnational migrations and existence in hostile environments.

*Juan Soldado:* Digital Print by Alma López, 1997
*Juan Soldado* is part of the 1848 Series on land, borders, borderless migrations, the
Treaty of Guadalupe Hidalgo, and Mexican Americans in the United States.

La Purísima Mission (Twentieth-Century Reconstruction)
Lompoc, California

# CHAPTER 10

# Spanish Caprice

Jesse Lerner
Rubén Ortiz-Torres

Soon may the Papagos gather
Beneath the sacred shade,
Where their fathers knelt 'round the Black-Robe
Listened, believed, and prayed.

Soon may the Black-Robe's labor
The treasures of faith unfold,
And this mission bloom in the valley,
As once it bloomed of old.

May its fading pictures be bright'ned
Its statues newly dressed
And the touch of the artist emblazoned
Its old Franciscan Crest.
May its arches again re-echo
The sound of the Vesper hymn,
And fervent souls to worship
Kneel in the shadow dim.
Brushed from each shrine and altar
The gathering dust and mold,
May the daily oblation be offered
Which the prophet hath foretold,
May its broken cross be uplifted,
And its bell more sweetly chime,
And its glory remain untarnished
Until the eve of time.

Ildenfonsus describing the mission
San Xavier del Bac, ca. 1919

AUTHORS' NOTE: This text first appeared in *Art Issues* #41 (January/February 1996), pp. 23-25. It is an adaptation of the narration from the second part of *Frontierland/Fronterilandia,* an experimental documentary funded by the Independent Television Service (ITVS) and the Fideocomiso para la Cultura Mexico-U.S.A.

Residence, Pueblo Revival
Santa Monica, California

Mission Inn
Riverside, California

For the North Americans who came to California during the second half of the nineteenth century, the Franciscan missions were nagging reminders that the West had not always been theirs. Ever since the liberal Mexican governments instituted policies of secularization, these distant outposts of a defeated empire had fallen toward ruin. But out from under these ruins grew an industry propagating the romance of old Spain—a fanciful vision of these buildings as picturesque relics from a noble past.

Where there is a noble past, or even the illusion of one, entrepreneurs, promoters, and clientele are on their way. Thus, although the origins of the late-nineteenth-century "mission fever" were literary, the missions nevertheless later inspired fiestas, parades, real estate development, and tourism. The missions also attracted the attention of architects and their employers. Rather than transplanting alien and often inappropriate architectural forms from elsewhere, they hoped to develop a distinctly Californian style of building, appropriate to the climate and evocative of their particular understanding of the region's history.

Although the original missions were designed as religious communities, these new Mission Revival buildings had other uses. Given this alteration of function, architects relied on the quotation of a series of evocative elements. Details that characterized the architecture of the original missions and that were para-

Movie Theater
Fullerton, California

Apartment Complex
Huntington Park, California

phrased by builders in the late nineteenth and twentieth centuries include massive walls of adobe (for which concrete and drywall were later substituted), the red tile roof, arcaded corridors, terraced bell towers, and the patio with fountain and garden.

Within a few years, Mission Revival had become the semi-official architecture of California. Architects built train stations, post offices, schools, airplane hangars, department stores, apartment buildings, bungalows, gas stations, presidential libraries, automobile clubs, and fast-food restaurants in this style. Endless permutations blended Mission with Craftsman, Queen Anne, Federal, and other diverse architectural styles. Mission elements were often mixed with or variously referred to as Spanish, Moorish, Romanesque, Oriental, Islamic, Latin, and Mediterranean styles.

Junípero Serra Museum
San Diego, California

California's Mission Revival proved to be only the first of a series of architectural styles that migrated across the border from south to north. The architect Bertram Goodhue instigated a vogue for the more ornamental Mexican *churrigueresco* style with his designs for the 1915 International Exposition in San Diego. Architects including Frank Lloyd Wright blended Aztec and Mayan elements with modernist forms, while others took those same pre-Columbian references in more flamboyant directions. Although many of these fads proved to be short-lived, the Mission Revival has remained the most lasting and characteristic architectural style of the California landscape. Beyond California, the taste for Mission Revival spread north to Vancouver's Chinatown, east to New England, south to Tijuana, and on to Mexico City,

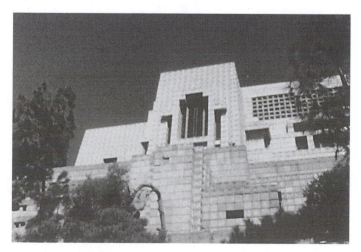

Ennis House
Los Angeles, California

Holt House
Redlands, California

University of Southern California
School of Medicine
Los Angeles, California

Mayan Theater
Los Angeles, California

where in the 1930s, there emerged a Mexican reinterpretation of a North American copy of a colonial Mexican architectural style.

In Mexico, the Mission Revival or *Colonial Californiano,* as it became known, referred less to the original missions than to the Hollywood dream. The buildings became more ornate, incorporating stained glass windows, elaborately carved stonework, and Baroque elements. Although modern Mexican architects disparaged the style as a kitschy, phony affectation of the nouveau riche, a revolutionary revisionism later came to advocate a style that was called Neocolonial Nationalism. The resulting buildings looked much like those of the *Colonial Californiano.* The early work of Carlos Obregón Santacilia, the leading architect of the Revolution, includes Neocolonial Nationalist housing for the workers, although in his writings he dismissed the California style as "pocho" (a slang word for someone who speaks neither Spanish nor English properly).

Ultimately, then, in reappropriating colonial architecture, both Neocolonial Nationalism and the *Colonial Californiano* emerged as something new. By the time a Mexican architect built a church in the Mission Revival style, it no longer looked like a mission. Mission Revival buildings, although they were always copies of something else, have subsequently been recognized as landmarks of architectural significance, both in Mexico and in the United States. Today, ironically, some of these buildings have been declared historical monuments, a status that they had aimed for at the beginning.

Throughout the twentieth century, the Mission Revival style influenced many important modernist architects working in California, especially Secessionists such as Irvin Gill and Francis Underhill. But the Mission Revival and modernism always made strange bedfellows. Anticipating later debates with postmodernism, the Mission Revival foreshadowed an interest in regional history as opposed to the development of a universal language—or international style—in architecture. Like the old Spanish Fiesta still celebrated today in Santa Barbara, the Mission Revival instigates a dialogue with the past that resonates in the present.

# CHAPTER 11

# On Foto-Novelas
## *Distant Waters* and *In the Mirror*

Carlos Avila

Executive producer Carlos Avila discusses *Foto-Novelas*, a four-part television anthology series that uses science fiction, fantasy, and magical realism to tell Latino stories. The series was presented by the Independent Television Service (ITVS) with funds from the Corporation for Public Broadcasting.

The idea for the *Foto-Novelas* series came from two comic book-style genres that are wildly popular in Mexico and Latin America: fotonovelas and historietas. As a boy, I remember being in people's homes in Mexico and in the United States and always coming across these picture books that were full of vivid imagery and used great storytelling techniques. The plots were usually high melodrama, with classic "rags-to-riches" situations or secretive, forbidden love between two persons from different social classes. Several of the stories had elements of the supernatural or were loaded with underworld figures that suggested film noir. There was also this strong idealism in the books—the notion that romantic dreams and aspirations can come true.

It's amazing how people respond to these books. Just walk down the street in Mexico City or in downtown Los Angeles, and you'll see massive amounts on sale at newsstands. In recent years, the historieta and fotonovela traditions have been appropriated to tell an even more eclectic variety of stories and to introduce new themes—the adventures of professional wrestlers, religious tracts, racy stories about truck drivers, pornography, social consciousness—but the great imagery remains, and their popularity is sustained. I think that the popularity of the form is what inspired me the most. Everybody reads them. Their popularity cuts across all social boundaries. So why not use the fotonovela and the historieta forms to tell stories about Latinos in the United States?

None of the episodes were overt adaptations of popular fotonovelas or historietas, but aspects of them did come into play. The fotonovelas and historietas that were my favorites had a dark and complex component to them—elements of the supernatural,

the magical, the darker side of life. But I was also interested in stories that had an idealism, which is very much a part of the genre. The stories I was looking for needed to blend these thematic and dramatic considerations. I wanted to steer clear of the more melodramatic side and try to present a more stripped-down, human story unadorned with excessive emotion.

The stories in *Foto-Novelas* are profoundly satisfying on an emotional level because the normal boundaries of our existence—time, locale, death—melt away to fulfill an unspoken desire: a family is united with its dead through a mirror, a young boy combines the love of his adopted parents with the missed things of his homeland, a man rediscovers the wonder of learning through a child. Fotonovelas engage their readers in a deeply emotional way. I think that their popularity is a tribute to that. But as with most comic books, the tendency is to work with emotions in an emblematic and overt way. It is part of the limitation of the form, although I think that graphic novels (which are increasingly popular worldwide) demonstrate that comic book stories can be told in a variety of creative and complex ways. The "classic" fotonovelas and historietas work on straightforward, clearly stated,

Images from *Distant Waters*[1]

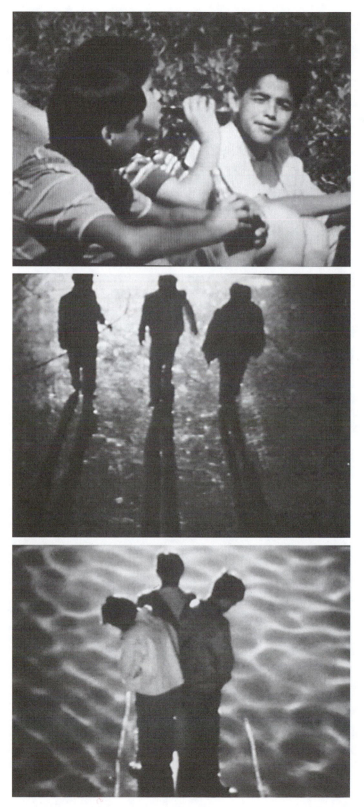

Images from *Distant Waters*[1]

emotional terrain. One of the goals of the *Foto-Novelas* series was to transcend the limited spectrum of Latino representations that we see over and over again—the gang-banger, the gardener, the downtrodden, the household maid, the sexy supervixen—and to give human characters unusual and creative circumstances in which to live. I was trying hard to move the drama into a different realm.

My concern is not that people will dismiss the series as action packed or simplistic but that they will disregard it because it has the veneer of being "just a Latino television series." I'm concerned that people will stick it in a box marked "ethnic" and never get past that label. The stories are human and universal, with emotions that everyone can tap into. They deal with the search for belonging, redemption, the quest for knowledge and self-worth, the need to overcome personal failure, and the human impulse to be the fullest expression of one's self. That said, they are also funny, entertaining, engaging, and visually striking. My hope is that people will see these stories as well-told contemporary human stories and not just these quaint, "ethnic" tales. The stories speak a universal truth and have an appeal that extends far beyond the Latino community.

The Latino population in the United States is much more diverse and complex than is usually reflected in the media. I think that cultural awareness is vital for anyone who is trying to carve out a sense of identity in a society that strives to homogenize for profit. The nearly 30 million Latinos who live in the United States come from so many countries and regions that it is critical to acknowledge their uniquenesses. A teenage immigrant from rural Mexico living in Vernon (Los Angeles) is different from a Peruvian professor living and teaching at a university in Philadelphia. Here in Los Angeles, I recently went to a big rock en español concert (essentially a hybrid form of rock that uses rap, funk, and traditional Latino musical forms and is sung in Spanish). I saw so many types of young Latinos. Looking across the faces in the audience, I couldn't help feel that this was America's future. It's clear that the dialogue isn't going to be only black and white in this country but that other people will have to be invited to the party. So much energy and culture are to be found within our communities; this music event was just one expression of it, and I hope that *Foto-Novelas* is another. Latino culture in the United States is expanding and exploding. It cannot be ignored.

And yet I'm cautiously optimistic about Latino representation in film and television.

Images from *In the Mirror*[2]

Although Latinos do not have a television series on the air that regularly features them as cast members or has a Latino setting, Latino actors are increasing in prominence, for example, Jimmy Smits, Roxann Dawson, and Cheech Marin. In addition, their roles are getting more complex and multidimensional. There is no reason why this shouldn't continue, but you never know—as creative a community as Hollywood purports itself to be, it never seems to be able to get beyond the well-worn stereotypes that find their way to the screen. The antidote to this is that there is also a growing body of Latino filmmakers. We're getting Latinos who are excellent as writers, producers, directors, cinematographers, and actors. We now have Latino "stars" who can be used as anchors for getting project financed. Things seem to be

poised for a future in which Latinos will find a place on screen. I put my biggest faith in the young Latinos working in film schools and independent films. The explosion in independent filmmaking

has empowered a lot of Latinos to work outside traditional financing methods and find ways to tell their stories. As Latinos get more skilled, the more the film and television production process becomes demystified. Different Latino stories will get told.

Images from *In the Mirror*[2]

## Notes

1. These images are from *Distant Waters* (1991), a short film written and directed by Carlos Avila for Echo Park Filmworks.

2. These images are from *In the Mirror* (1997), directed by Carlos Avila. *In the Mirror* is an episode of the *Foto-Novelas* anthology series produced by Echo Park Filmworks for the Independent Television Service (ITVS) with funds from the Public Broadcasting Corporation.

# CHAPTER 12

# Poems de una Chicana/Irish, Chuca Veterana que piensa que es Sirena turned Activist, y Poeta para acabarla de chingar
or
# Mermaid at the Edge

Lindsey Haley

Al rato nos watchamos

Yo con mis Levis muy planchaditos
Usted con su Pendelton muy chingón

We made a pretty pair us two.

"Angel Baby" was my song
"My Guy" was for you

We had visions and dreams

We'll have a little house,
picket fence in front,
nopales growing out of old painted yantas
and plastic ducks out on the yard too.

A little girl for me
A little boy for you.

I would get panzona,
You would get pelón.

But we promised to love each other true,
even at sixty-two.

Where did we let these dreams go wrong?
Did growing up come too soon?

Questions with too many answers,
yet never the right one.

Look at the monsters we've become.
Leaving us with only one thing left to say:
"Al rato nos watchamos"

■

Untitled

What can I possibly say now in your silence,
that while still living you could not hear.

We were never given that "last moment"
and so now my poetry will have to do.

Heading south on Highway 99
while the soil above you was still moist,

I thought I would never see you again
But I do.

and will for the rest of my life
through

His smile
Her eyes.

I see you everyday
I put you to bed each night

and

I love you all over again.

■

The Whisper

Regardless of what I may do,
What accomplishments
I have made

I am still belittled
by your ignorance

I still remind you

of your maid
the gardener

the gang member
the cute little prostitute you had on your last trip to Mexico.

As I hold on to what little I have left of my culture
you handcuff me and nail me to the wall,

The Whisper getting louder . . .

"If they love Mexico so much, then why don't they go back!"

It does not know that Mexico doesn't want me.
I am and am not one of Them.

I am the hopeful child.

I am the one my great-grandmother carried across the river
one hundred years ago.

We are four generations of strong women,
My great-grandmother
grandmother
mother
and
I.

Nevertheless, we were cut by your tongue.
With each passing generation
we bastardized our Spanish to the point of ridicule.

At the age of ten
I perfected my english
and write in its letters.

I don't pick grapes anymore
or make tortillas before dawn.

I dress in high heels, skirts
am well read
and write my own version of poetry.

And even though words may not say it,
Your eyes do.

The amazement that we
(your maid

the gardener
the gang member
the cute little prostitute you had on your last trip to Mexico)

Could possibly know as much,
if not more than
you
Could be just as
creative
sensitive
living
as You.

We are not your machines.

We all do not clean your toilets
We all do not look after your manicured lawns
We all do not fall trap to our country's ambitious
cocaine venture
We all do not provide you with a little bit of
hot brown ass.

Every time a maid
gets yelled at for not taking the message right,
Every time a gardener
can't go out to earn a day's living because it's rained,
Every time a thirteen-year-old boy
tries to sell me his mother's wedding rings,
Every time a prostitute
gives us her spirit to turn a trick,

I die inside.

The Whisper remains,
getting louder,

As all our children
Yours
and Mine

Continue to see us as
the maids
the gardeners

the gang members
and
the prostitutes.

■

Q-vo José (te manda saludos    You spoke caló
La Lindsey)    and

I remembered frying huevos on the patio on hot summer
afternoons
and playing in my Tío Chano's junkyard, adobe houses,
unpaved streets, dirt beneath my feet.

Remembered what I had forgotten

Tías chismosas, tattooed tíos,
baby showers, birthday parties and 4th of July's where the men
hung around the barril, and the women looked like
Sophia Loren
and smelled of Aqua Net.
Sunny and the Sunliners blasting out of a console stereo
standing on its skinny legs
while me and my primos played el candado out on the street.

El Chuco is a place far from L.A. I remembered,
as I heard your singsong rhythmic voice,
my memories bouncing along.

I remembered being pulled up from my roots,
transplanted in Califas, concrete beneath my feet.
Becoming a Chuca Sureña style

Years have passed,
I let my eyebrows grow back and have long since retired the
false eyelashes
and
it's been too long since I have spoken caló.

I wondered if my primos and their children still speak caló,
or worse,
If they speak perfect english.

I wondered as I stared at my beer, listening to the blues band
blasting in the bar,

Then I remembered
a Sunday
Not too long ago
Staring into the faces of 200 Cholos and Cholas at a peace
treaty picnic
where I didn't hear any caló
Not even an "Al rato te watcho," "chante," "refín," ni "carnal."

We heard speeches of how the freeways had divided us, how
barrio warfare had to stop.
and
I wondered
if my primos and their children,
some from San Juan, others from Tejas Edition,
still carried on the rivalry con El Segundo Barrio.
Wondered if they killed each other too.

I heard my father; a veterano, and the finest Pachuco
in my eyes,
mutter, "Son una bola de putos"
A proud man, whose time had required that he earn his respect
fighting con mano limpia
witness the decay of his grandson's generation.
Le cala, me cala, nos cala
Y aquí ya nadie habla caló.

# CHAPTER 13

# Pico-Union Vignettes

Luis Alfaro

**On a Street Corner**

So, this man and this woman are walking down Broadway in downtown Los Angeles. The man looks at the woman and says, *Bitch, shut up.*

The woman looks at the man and says, *Aw honey, you know I love you. I just wish you wouldn't hit me so hard.*

The man looks at the woman and says, *What? What? You want me to leave you or what?*

The woman looks at the man and says, *Aw no baby. You're the only thing that I remember.*

Desire is memory, and I crave it like one of the *born-agains* in my mama's church. But it's hard to be honest sometimes because I live in the shadow of the Hollywood sign. Because I live in the same town with the people that bring you *Melrose Place* (my weekly dose of reality). Because on a street corner known as Pico and Union, my father made extra money on pool tables, my mother prayed on her knees.

A woman danced in the projects across the street. I could hear the sounds of a *salsa* song as her hips swayed. Each step got bigger and bigger as she thrust out her elbows and clenched her fists. Her husband would beat the shit out of her with large big hands that looked like hammers. Each blow penetrated her face like a slow-motion driver training film.

A drunk from the bar at the corner staggers home, pushing people aside like a politician working a convention.

A man on the Pico bus gets slapped by this woman after she sat on his hand in the seat next to his. He says, *Hey! If you don't like it, don't sit here.*

On Tenth and Union, I forced my first kiss on Sonia López in third grade. The slap she gave me felt so good, it must have been my introduction into S&M.

A glue sniffer on Venice Boulevard watches the world in slow motion.

*Bozo the Clown* was at the May Company on Broadway. He's throwing out these gifts to all the kids. We're all waving and screaming, hoping to catch one. He throws this Monopoly game out at this little boy and it hits him right above the eye and he topples over. He comes up screaming, crying and bleeding and I watch in horror, afraid that *Bozo the Clown* will throw something at me.

People in this city used to run at the sight of a helicopter light. Afraid that their sins would show through like the partition at confession.

An earthquake shook and our neighbor is running down Pico screaming that *Jesus has come back, just like he promised!*

A man got slapped.

A woman got slugged.

A clown threw toys.

A drunk staggered.

An earthquake shook.

A slap.

A slug.

A shove.

A kick.

A kiss.

■

The Three Mexican Musketeers

The Three Mexican Musketeers. That's what we were. Small-time change in this big slot machine of a city. I worshiped my brother, Jaime, like little kids worshiped Cassius Clay or Lew Alcindor. Loved him because he always opened the back door of the bus so that I could sneak on. The third Musketeer was our neighbor, Gabriel, whose dad was gone by the time we longed to go to ball games with the absent guys they called *father*.

We befriended Gabriel after his dog Brandy died. We felt sorry and helped dig out a grave in his backyard on Valencia Street. Stole plywood from the Pico-Union Projects construction site and made a big cross like we had seen in a *stations-of-the-cross* Tijuana trip. At the funeral, my brother gave him our favorite 45-rpm single to play on the portable record player we bought at Zody's. He hugged Gabriel like a father when he started crying in the backyard as we put up that big cross over his buried dog, while the portable wailed, *Ooh Brandy, you're a fine girl. What a good wife you would be.*

After that, Gabriel was our friend, even though he ate non-Mexican food like *waffles* and *crepes*. But we had things in common. He was fatherless to a long-ago divorce, and we were fatherless to a lost-cause dad who spent his days at the *Cantina Jalisco* on Union Street. We were a gang of misfits. Different from the Eighteenth Street boys, 'cause we were nice and we didn't sniff no glue. We played out our adventures on the streets of downtown Los Angeles before the graysuits left all the buildings on Broadway.

Downtown was our backyard. *El centro* was the place to run through education electives that weren't offered at Tenth Street Elementary School. Stuff like *Heavy Petting 101* or *Advanced Shoplifting* on Olvera Street.

That summer we almost lost my brother to a girl who lived in the *Maravilla Projects,* who Gabriel and I thought was, like, just *all right.* Her name was Lorraine and the only thing cool about her was her *chola* eyebrows, which were shaped like Elizabeth Taylor's in *Cleopatra.* Luckily, my brother got bored with the bus trips through East L.A. and returned to our Musketeer adventures downtown.

One Saturday, we scoured the neighborhood for soda bottles. Turned them in for the five-cent deposit so we could pay for the bus fare and a kid's admission to the movies. We stood in front of the Tower Theater at Seventh and Broadway looking for somebody who would pretend to be our parent and buy us tickets for an R-rated *Kung-Fu* movie.

All of a sudden, Gabriel started to cry. Not just cry, but sob. Like he saw a devil, saw a ghost. We walked around the corner, down near the racing track form stand. My brother pulled up his own T-shirt to wipe Gabriel's face and held his hand. For a moment they looked like those pictures of little Madeline and her friends in Paris.

*I saw my dad.*

That's what he said, just like that. Gabriel said he saw his dad drive by in front of the Tower Theater and wave at him. I just didn't get it. I told him that if I had seen my dad driving by, I would've run up and gotten in that car and made him drive me home instead of having to get on that rowdy No. 26 bus where somebody always plays their Marvin Gaye way too loud.

We stood in front of the Tower Theater and waited for Gabriel's dad forever. Which in kids' time, is really long. We missed the first showing of *The Chinese Connection.* I prayed we wouldn't be late for *Enter the Dragon.* The thing about Chinese *Kung-Fu* movies is that if you miss the first five minutes, you miss the whole reason they spend the rest of the movie fighting. Besides, we couldn't rush Gabriel, who was waiting for another chance to see his long-lost dad. Somehow, we all knew what that meant.

Right when it seemed we were going to die because the next set of previews was already on, a green station wagon pulled up in front of the theater. Sitting alone in the front seat was Gabriel's dad who smiled all nice like he'd been gone only a few minutes. Like instead of being gone forever, he had just gone out and bought milk for the kids or something.

Gabriel didn't even say good-bye. Didn't even give us his movie entrance money or nothing. He just hopped in the car and drove off like he'd been waiting all his life in front of the Tower Theater for his dad to just drive by and pick him up.

And that was the last we saw of Gabriel. We went in and watched all of *Enter the Dragon*, not having a clue as to why Bruce Lee was angry and who we were supposed to be cheering for.

Later that night, Gabriel's mom came over asking where he was. My mom brought her into our room, and she looked terrified. She was wearing her *tamale* apron with her hair up in curlers. When my brother told her that he went with his dad, she fell down sobbing and screaming. She went crazy right in front of us. My mother and father tried to pick her up off the floor as she screamed that *he'd stolen her son*.

I told my mom to call the police, but no one did. In our neighborhood, no one ever calls the police. Sometimes you call the ambulance or the morgue, but never the police. There are too many *ilegales* on our street, my dad included, to have the police snooping around our block looking for *wetbacks*. Besides, by the time she came over that night, Gabriel and his dad were way past the border and deep into the night on a Mexican highway. And that was the night that the voice went out of Gabriel's mom. She never spoke again. At least not in the neighborhood.

Gabriel returned home the year we started seventh grade, but there was no more room for the Third Musketeer. We were eating up the American Apple Pie like there was no tomorrow. He was *Mr. T.J., Mr. Taco* to us. Came back speaking Spanish and wearing *huaraches*. We were strutting through junior high in our *Superfly* platforms, talking *Soul Train* dialects. We didn't have time for no Third Musketeer. Once, I saw Gabriel at a bus stop on the way to school. In true junior high fashion, I ignored him, even though he stood just a few feet from me. I was into noon dances and Barry White, while he was hanging out with José Chávez and the Soccer Club.

But I'll tell you one thing. Even though we'd never admit it, the remaining Two Musketeers spent about six months after that incident hanging out in front of the Tower Theater. Hoping that a green station wagon was cruising down Broadway looking for us. We spent more Saturday afternoons standing on Seventh Street with stolen cigarettes dangling from our mouths praying for an absent father who might be looking for a couple of Musketeers on a street corner.

I would have missed the previews for that one.

■

Ama    What I remember is her skin. *Piel.* Wide and soft like those sheets that come out of the maxi-heat dryers at the *lavandería* on Virgil. Her body, a *Mexicana* glaze, color of brown sugar, evenly spread across the crevices. A Third World mark on the arm from an early inoculation that brands her a member of a tribe of *mujeres, las* who have crossed borders, worked farmland, sewed, stitched, cleaned, and *whatevered.*

*Yeah, whatever, because I am not my trabajo,* she says. *I am not my trabajo.*

Her body has already started to go through *La* Change, but she's still working the Ava Gardner. On Saturdays, we go watch dad play soccer at Hollenbeck Park, and she is working the Black Widow skirt and the Elizabeth Taylor cleavage. Sundays we make entrances at our Parish, Immaculate Conception, like we are walking into La Scala. Pico-Union mothers always work the *Vanidades* cover look, walking the *barrio* runway like *Dolores Del Rio* clones.

Then there is the night. Friday night, paycheck night. The smooth *tequila* poured slow into a shot glass next to a small cup with tomato juice, Tabasco, lemon juice, and chopped onions. The chaser and a kiss. My father trying to be *Francisco* Sinatra but looking more like *El* Sammy Davis Jr. Waltzing in the kitchen with *ama* in her *María Félix* finest.

Late night Fridays after ten o'clock. After *Love American Style.* After bedtime for the kids. After the last light goes out at *El Parían Birriaria* and all chances on a late night *taco* run are lost. A scratchy *Javier Solis* on the turntable accompanies my mother's giggle and the shuffling of dancing feet on the brand-new tile from Sears. My drunk *apa y ama* accompany the recording in a slurry duet, followed by soft in-the-dark whispers about money, work, schoolbooks, and the Beverly Hills boss who wants us to pick the avocados off his tree. But it's not all that bad because all that *guacamole* we pick off the Camden Street tree means *taquito* Saturdays and *oh so gratefuls, oh so gratefuls.* We escape into the station wagon and breathe easier as Wilshire becomes more familiar past Western, before Bixel.

Her body a map. Each fold, curve, wrinkle bearing witness to city life. To the daily hanging of laundry on a clothesline that stretches to the skyscrapers down the street. To Delano days in the hot sun and those grapes, those goddamn *uvas,* and her back, *la espalda y esas pinches uvas.* The body tracing experience with a scratch, a scar, a mark on the skin for the moment when it all went bad. When *apa* took the time-outs, waking up in alleys. Dealing with the workweek and the stress with a new late night lover, *la botella* at *El Club Jalisco* on Vermont. Then, and only then, could we stay up late and watch Hans-something-or-other on *Fractured Flickers*. Or *El Panzón*, Jackie Gleason *y Los* Honeymooners. But all I can remember in the late night are the opening credits and that moon face smiling down on an apartment building in New York. And my mom's on the couch looking out the big picture window in the living room waiting for him. *El* Sammy Davis Jr. to waltz in with a bag of *birria tacos* and a six-pack of some bottles from a Mexico manufacturer, all fake yellow, something *piña,* cold and wet.

I know that smiling moon on late night television lies, lies, lies. Because I've stayed up late and watched *El* Jackie Gleason scream at her *y* throw tantrums. *El* Jackie Gleason threatening to leave. *El* Jackie Gleason threatening to go out and get something better. And it's funny 'cause he never does.

So what happens when *El* Jackie Gleason winds up at some *cantina* in downtown Los Angeles with the Friday paycheck? He does a quick five shots of something or other and buys everybody a round. He bets all his money on a bad luck pool table game and pushes somebody around in his drunkenness and gets slugged and ends up on a barroom floor. He chases it all down with too many beers on a credit line that he will never be able to pay for. When the sun comes up on his empty pockets, all he can think about is the time clock and the daily grind and the Vernon factory line and the production quota and those carburetors. *Esos pinches carburadores,* like those *pinches uvas.*

Somewhere in the middle of all that thinking, he blacks out and wakes up in some Pico-Union alley where his two little boys recognize him and carry him home. The Ava Gardner outfit gets thrown out, like an old Halloween costume, into the metal trash cans with the cool aluminum tops that always make great shields for the Roman warrior battles we had with the boys from Valencia Street. And she looks like *La* Audrey Meadows in a robe we bought her one Christmas from Target back when it used to be called Two Guys. I remember it 'cause it was in Alhambra, backyard of *El Sereno,* where you could buy a hot dog for thirty cents. This old African American man sold the plump ones and always called my *ama,* mam. And boy did my *ama* sure look fine to him. Like a movie star. Like Tippi Hedren of The Birds. That's what he said. *Except your mama's all brown, but that's all right 'cause your mama done worked for it. She earned it. Uh huh. Down there in Mexico, you got your queens and you got your women gods and they all better than that ole queen in England 'cause Mexico queens they earned it. Yes, your mama looks just like Tippi Hedren of The Birds, except she brown. Lord.*

And my *ama* is honking the horn while she pulls up the car to the curb to pick us up from the hot dog stand where she dismisses that crazy old man who takes advantage of a woman. Knowing that just one week ago those hot dogs were twenty-five cents, and all of a sudden he sees a lady with four children and no husband . . . *sin esposo*.

A soft sniffle is her chorus home as we drive down Huntington Drive through Lincoln Heights. Past General Hospital where we don't even stop at the McDonald's on Marengo. We keep going down Macy, down Sunset, down Figueroa. We pass a line of the same ten people at The Pantry. A quick zoom through the Union 76 ball that turns and turns. On Pico you can hear the sound of relax in a quiet sigh she makes as we pull into the driveway and notice that, once again, there is no *him*.

And somehow, I'm back at the end of *Los* Honeymooners, except I don't know it 'cause I never make it to the end. I know the moon is warm and runny and that her tears fall at the end of *Los* Honeymooners as she picks me up and puts me into a bunk bed in the middle of a dream sequence where he, *El* Sammy Davis Jr., is fighting off lions, dragons, a Beverly Hills boss, and the stress. *Ese pinche stress.*

She is walking through the house in her black *falda* and bra. Asks me to zip up her muumuu from the back while she ties the handy tortilla-making apron around her waist. Then I see it. The back. A map, bearing witness. *Una espalda de testimonio.*

I see it in her back. I see *la frontera*. I see my grandparents crossing over. I see the dusty Delano streets and those grapevines. I see the peaches ready for canning and the lining up of the Kerns jars. I see her serving the Root Beer to *los* Pilipinos at the A&W with the Papa, Mama, and Baby burgers. I see us kids on the porch with yet another family dog that will get run over on a downtown crosswalk. I see the clotheslines. I see the comal. The chile-making bowls. The domestic duties.

And I see him. Waltzing into the kitchen after *Love American Style*. With the shot glass. With the *Tres Flores* in *el pelo*. With the Sammy Davis Jr. strut.

And I see her. The brown-sugar *Mexicana* skin. All Ava Gardner. All *María Félix*. All Tippi Hedren of The Birds. Working her brownness while plucking a downtown chicken. *Chisme* on the *teléfono* from the grapevine down to Tijuana. Waiting for him *en la sala* at the big *ventana*. The dress. The apron. The Audrey Meadow-ness of her.

I see it in her back. *La espalda. Lo que me enseña.* She has earned it.

She has earned it.

# CHAPTER 14

# L.A. Cucaracha Urban Sketch Journal

Lalo Alcaraz

The landscape of Los Angeles has long served as a generous muse in my editorial cartoon work. She may not always be pretty, but she's always willing to inspire a 'toon, or stand back and let me scribble on her as a backdrop to the ironies and foibles of the Latino cultural and political landscape. My strip began to run in the *L.A. Weekly* in June of 1992, just weeks after the infamous Los Angeles riotous uprising of April. The first panel was #1, "White Men Can't Run the System," a parody of the film *White Men Can't Jump* and a commentary on the lack of control and leadership from the local and national power structure.

#2, "(Help the Authorities) Find Cuco Now," casts the main character in my strip, Cuco Rocha, a disaffected Chicano cockroach, as the object of a Find Waldo parody. The setting is recast as somewhere in bustling Boyle Heights and restates the Rodney King beating with a mostly Chicano cast.

#3, "Malti-Cultural," comments on some of the reasons L.A.'s inner city may have been so quick to go up in flames—cheap liquor targeted at minority imbibers.

#4, "Hell Taco," spoofs some of what passes for Mexican food in Los Angeles, the world's second largest Mexican city, namely, extremely processed Americanized fast-food chain food.

#5, "Hidden Banks." I find that the mainstream media often covers only white middle-class concerns and leaves out many of the issues important to inner-city minority dwellers. The lack of convenient and affordable banking facilities in the inner city was ignored during coverage on the increase of automated teller machine fees.

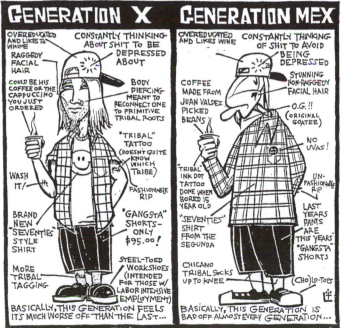

#6, "Generation X/Generation Mex." Los Angeles is a cultural capital for mainstream culture, but many don't acknowledge or even realize that out of Los Angeles flows much of Chicano culture and style. These Raza stylings are then spotted and imitated by white/Anglo youth and then commodified by American capitalist culture.

#7, "Alatin." Although Los Angeles is a majority minority city, its Latino majority is not adequately represented in the entertainment industry. Because of the scarcity of nonwhite executives, the movie and television industry is acquainted with Latinos only through their major roles in the service industry of Los Angeles, that is, valet parkers, cooks, busboys, maids, nannies, and so on. This may give them a skewed view of what roles Latinos play in the real world outside their rarefied Hollywood confines. I've basically been forced to create my own Latino movies in my cartoon, without necessarily imbuing the mall with a strong political message, although . . .

#8, "Pocha-hontas," comments strongly on the lack of Chicanas' identification with their indigenous heritage and on the paucity of images to cull from. I was left with only Disney's Pocahontas to relate to many Chicanas' strong native features. The original inspiration came from a poster I observed in an East L.A. football game, the Fall Classic, also known as the Chile Bowl. The game pits two heavily Chicano high schools, Roosevelt and Garfield. The Garfield junior varsity cheer squad had a large poster featuring Disney's historically inaccurate Pocahontas cartoon character in a Garfield cheer sweater. On a certain level, they realized that Pocahontas, stylized as she may be, had more than a fleeting resemblance to themselves.

#9, "Millions of Juans." This movie poster parody is based on *Million to Juan,* a forgettable movie made by comic Paul Rodríguez, in which he played one of L.A.'s ubiquitous orange sellers who receives a mysterious financial windfall. I wanted to get across a message of political strength in numbers, not necessarily the kind of numbers you get from winning the Lotto.

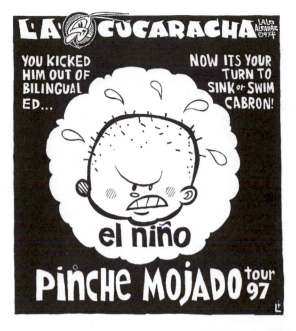

#10, "El Niño." Much of the political tension generated by vote-grabbing politicians in California has been generated by a pathological demonization of Latino children. They were scapegoated in the anti-immigrant schoolchildren voter initiative Proposition 187, during the anti-affirmative action Proposition 209, and during the recent anti-bilingual education "English for the Children" voter initiative. Even the weather pattern ironically known as El Niño (the Child) has been blamed for every weather disturbance under the sun. It must be easier on one's conscience to blame an immigrant child for one's meteorological problems than to blame, say, global warming.

#11, "Elementary 187." Thanks to the divisive Proposition 187 campaign of 1994, the political landscape threatened to change into a Nazi-era atmosphere of ID checkpoints and fences designed to keep immigrant children out of school.

#13, "Separatism in Quebec." How ironic I found it that while many decried the separatist movement in Canada, led by Quebecers, whites in Los Angeles kept moving away from their old neighborhoods as soon as minorities moved in. Los Angeles is dotted with exclusive gated communities, which are America's great contribution to separatism.

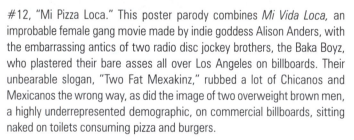

#12, "Mi Pizza Loca." This poster parody combines *Mi Vida Loca,* an improbable female gang movie made by indie goddess Alison Anders, with the embarrassing antics of two radio disc jockey brothers, the Baka Boyz, who plastered their bare asses all over Los Angeles on billboards. Their unbearable slogan, "Two Fat Mexakinz," rubbed a lot of Chicanos and Mexicanos the wrong way, as did the image of two overweight brown men, a highly underrepresented demographic, on commercial billboards, sitting naked on toilets consuming pizza and burgers.

#14, "Separatist!" Often, after segregating minorities to their own neighborhoods, many Anglos then turn around and astoundingly accuse those same minorities of separatism, because they "only want to be with their own kind."

#15, "Café Attitudo." Living on the east side of L.A.'s vast metropolitan sprawl, I don't usually venture all that far into West L.A., but when I do, I often have odd encounters with the strange foreign culture of the more affluent and chronically snide Westside. This fellow actually "happened" to me and a few Chicano friends at one of your snootier establishments.

#16, "Wacky Wilson's Welfare Warehouse." California Governor Pete Wilson, in addition to scapegoating immigrant Latinos and adding to their stress and paranoia, also turned his aim at the poor in general. Los Angeles County bore the brunt of Wilson's political scheme to further disempower minority poor.

#17, "Infosuperhighway." Freeways in Los Angeles tend to run right through the middle of many Chicano/Latino neighborhoods, but the new information superhighway seems to be taking pains to avoid disturbing any lower-income barrios.

#18, "Flying Quake House." Earthquakes are part of L.A.'s character, and they generate almost as many complaints as the presence of Mexicans do. I have long wished that people who complain constantly about L.A.'s faults would just act on their whining and take the first flight out of town, preferably to the much vaunted "Middle America" everybody keeps talking about.

#19, "If I Designed the Getty." Before I was an urbane professional cartoonist, I received an M.Arch. from Berkeley, which I feel qualifies me to completely redesign L.A.'s newest landmark, the Getty Center, with my own bent Chicano twist.

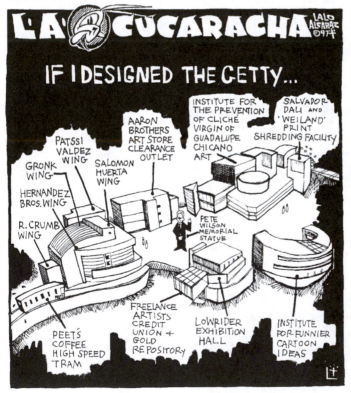

#20, "Che 1997." Can I just say that the Nike swoosh is on practically every young Latino youth in Los Angeles, and I am just sick of seeing it? This cartoon represents to me the world's condition today: the insidious commodification of everything and the infiltration of capitalism into political progress movements. By the way, this is for sale as a print. . . .

# CHAPTER 15

# L.A. Real

Theresa Chavez

**Introduction: First Performance**

In third grade, I came home from school with a book titled *The History of Monterey Park*. When I showed it to my mother, she declared that the sullen-looking gentleman pictured opposite the title page was my relative. She knew his name but wasn't quite sure of the exact lineage—some great, great, great, great-grandfather of mine. In 1810, Don Antonio María Lugo received a permit from the Spanish government to graze cattle on 29,000 acres. It would later be referred to as a land grant, encompassing a number of Los Angeles-area cities including what is now the working-class suburb of Monterey Park. The complete meaning of that knowledge was beyond my grasp, but the moment of receiving that information was an exciting one. My mother's pride shone through as she, for the first time in my short life, explained the "story" of our family. The "story," as I was later to discover, was an oral history full of misinformation and nostalgia for a culture that barely exists today.

I returned to my third-grade classroom and explained to the teacher my unbelievable news. I was related to one of the founding families of Los Angeles. His picture is in our third-grade reader. She too was amazed and asked if I would make a presentation to all the third-grade classes. I agreed, and soon after—in a blue velvet dress that my mother had sewn for the occasion—I delivered a short explanation of a history as told by my mother. My first public performance of my family's history was a success. The students and teachers were impressed and listened attentively. I remember holding notes in case I forgot an important detail. I can still see my writing on those yellow and green index cards—soldier, rancho, King of Spain, cattle, adobe, taxes. . . . Through the act of public presentation, I had personally accepted the weight of my own history. From that day forward, I knew that the story would have to be told and retold. I had no idea what form those stories would take, but I embraced the familial responsibility of being the keeper and caretaker of our past.

The following one-person performance script is one form that this story has taken. It has been performed for numerous audiences throughout Southern California as well as in New York City.

L.A. Real

*The Mestiza Narrator, a contemporary woman of Indian and Spanish heritage, is sitting in a chair, stage center, amid a pile of books, papers, and paintings, including a mid-nineteenth century diseño (map) of Los Angeles (mural size).*

*Mestiza (taped voice-over):* Can you imagine a complete quiet and stillness—only the sound of the wind and an owl? But you are standing on what will become the corner of Spring and Third Street downtown. You see just the land. The earth is dominating your view. The earth in front of you, around you, is so big. It is so much. You must listen to what it has to say. And it is talking all the time. You listen between bites of food. You listen as entertainment. It is as present as our garbage truck television hammering sound. You don't know what "getting away from it all" means. So you hear and then you speak with such solemn respect and hope, to convey your heart as when you must let your lover know in some way what you're feeling—that deep saturated emotion. It is to tell them, to connect with them because something is happening between the two of you— between you and the earth. You hope you are loved back. And you listen, and you know you are. But you are standing on what is now the corner of Franklin and Vermont, facing north toward the hill. A hill where there is no Hollywood sign. Imagine. You're walking. After one hour, you get there. You

Ramona Lubo, ca. 1899, whose life inspired the best seller by Helen Hunt Jackson and numerous films.

stop, look at the hill where, if you climb it . . . you see an ocean. But I don't. Today I can't. I have a meeting. A council. Some kind of alien has arrived. Is he a time traveler? Has he broken the sound barrier? How did he get here? He wants to live here. He is not a listener. When he comes near me, I hear the garbage truck television hammering sound. But it's really only his armor. He calls it armadura.

*Slide: Contemporary Los Angeles*

*Mestiza:* In 1771, before this place was America the United States, before this was Mexico, my family came to live here. This was then El Norte, la tierra al dentro, las provincias internas. They came from Sinaloa and Sonora. Los pobladores, the settlers—Africans, Indians, Spaniards, Mulattoes—that's half African and half Spanish, Mestizos—that's half Indian and half Spanish, Coyotes—that's half Mestizo and half Indian, and one Asian, probably Filipino. They were separated from the rest of New Spain by the vast Mojave Desert. Spain becomes a figment of their imagination. They are now in a new place, separate and willing to redefine themselves according to the whims of this new earth and, of course, the whims of the Spanish government. Spain is now only a fragment of my body. It breathes alongside the Indian in me. This Mestiza mix-up in me feels the Middle East, too. . . .

*Slide: Lugo family, ca. 1890*

*Mestiza:* I could just as easily pretend that none of this happened, that I am not related to this past. But I still find traces of it in my body. They sit alongside the mercury and fluorocarbons I can trace as well. It would be easy to forget amid the strangle of freeway around my neck—forget the complicated past that covers seven generations, three federal governments, a land grant that would now cover six different cities, the breaking of the 1848 Treaty of Guadalupe Hidalgo, a barter economy turned cash economy, a crooked land commission, real estate promotions, and the attempt to undo the use of all indigenous languages. I could go about my business and forget. But I see pictures. I hear stories. I read the printed word. A tribe of my people lived here. But whatever it is they created, it no longer exists. Their songs do not get sung. Their land has been cemented over. Their paths have become our boulevards. I can barely see the world that was theirs. Why should I remember a past that only complicates my living, my understanding of who I am? I could simply selectively edit my own history. Give my own face a new meaning. Be an American mongrel. Or define myself according to any given historical moment—Californiana, Mexicana, Mexican American, Chicana, Hispanic, Latina, Mestiza . . . Californiana . . .

*Mestiza picks up painting of Antonio María Lugo, places it on an easel.*

*Mestiza:* This story begins with the Californios. As a first-generation settler, my fourth great-grandfather called himself this. As does my mother, as did her father. As I can. It's just that no one understands what I'm talking about. I tell them it makes me feel as if I, personally, have been walking around here for more than 200 years. Watching el río loco, the L.A. River, change course at whim. Watching the drought of 1862 bake the earth and drop the final ax on a culture being eaten up by manifest destiny.

My great, great, great, great-grandfather, Francisco Lugo, was the first to arrive in this so-called "land within." He was a soldier in the Spanish army, stationed at the San Antonio Mission, near San Francisco. In 1810, his son, this man [*points to painting*], Antonio María Lugo, at the age of thirty-four, applied to the Spanish government for a grazing permit to graze cattle. He received 29,000 acres.

His family's cattle grazed on what are now six cities south and east of downtown—Lynwood, Huntington Park, Bell, Maywood, Montebello, and Monterey Park. Cattle was a product industry then. Cattle hides were used for trading in the international hide and tallow industry. Cattle raising did not become a meat industry until the Americans arrived forty years later. In 1830, there were 2,500 sheep, 3,000 horses and mules, 43,000 head of cattle, and 4,000 persons living in L.A.

*Slide: Los Angeles Pueblo, 1860*

*Mestiza:* Antonio built a home on the Plaza, the Pueblo's place of genesis. After the family gave this house to the church,

*Slide: Lugo house, ca. 1910*

*Mestiza:* They built this home in 1860 in what is now the city of Bell, south of downtown, where the family lived until it was sold in 1981. Several months before it was officially designated as a state historical site, it was vandalized and burned to the ground.

*Slide: Lugo house as Chinese restaurant*

*Mestiza:* The house on the Plaza later became a center for Chinese social life during the Plaza's Chinatown years, in the 1920s, before Chinatown was "relocated" to create a tourist area called Olvera Street. The city fathers decided "it is better business to get off the train at Union Station and see piñatas instead of plucked ducks."

*Slide out*

*Mestiza:* By 1900, only fourteen acres of the rancho were left. The family lived in Bell with a few horses, one of which my oldest cousin, Laurita, used as transportation until she was about twenty. She is now ninety-two—still amazingly lucid and still attached to this earth and how her family lived on it. I remember several years ago, when I first walked into her home, how she embraced me. As she pulled me to her chest, I felt as if she was pulling me backward, taking me back in time. Everything felt so different. She felt different. Different chemically, as if her body was made up of a different kind of oxygen, a different kind of $H_2O$ that had poured into her body in a way that it had not poured into mine. Time passes differently through her veins. Her eyes speak wide-open landscapes. Like her mother, she still keeps the dinner table set in case someone might drop by. In her mother's time, the door would be opened to anyone the family might know or to those who happened by and needed to water their horses. Dinner was always ready.

Lugo family in front of their adobe home in Bell, California, ca. 1890.

Now, Laurita's table is set as a memory, like an altar in honor of all the dead who passed through the family homes, in honor of all her lost rituals.

*Slide: Bell house, ca. 1890*

Of the three houses that Antonio and his family lived in, none remain. The Lugo house on the Plaza was torn down in the 1950s to build the Hollywood Freeway, even though the house was not in its actual path, standing several hundred feet away from the highway. I never lived in these homes. I visited one once—the one in Bell. Unbeknownst to me, the property was being prepared to change hands, from the Lugos to an industrial park visionary. An older, distant cousin was living there. I didn't even have time to introduce myself before I was shooed off the property.

"Quítate de allí. I'll call the police." She is frightened by the sight of a stranger standing too close to her front door. "I'm just looking at the historical marker, trying to verify where I am, verify who I am." The house is now surrounded by an industrial suburb of downtown L.A. Right outside her front door is a major boulevard with rush hour traffic. The house that once stood near the center of 29,000 acres has become just another structure mixed in with storefronts, Taco Bells, and major home appliance factories. I think this is the first time in quite awhile that she has even stuck her head out the front door. Inside it is 1901. Outside it is 1980. Outside it is English. Inside it is español. Inside it is wood. Outside it is asphalt. Outside there is time. Inside there is space. Inside her is resolve. Inside me is conflict. I feel a part of this place, but I have no business here. This is my past, but I have no future here. Maybe she had accepted her losses. But when the house burned down a year later, I could not accept it. I went looking for anything that might be left—stories to pass on.

> *Music and text slides up: The story . . . the teller . . . the truth . . . the details . . . the rumors . . . the opinions . . . the basic lies . . . the angle . . . the light . . . the version . . . the distance from the event . . . the tale of one city . . . the romance . . . the legends . . . the sequel . . . the saga . . . the reporter . . . the portrait . . . the transcription . . . the frame . . . the family name . . . the family lie . . . the translation . . . the remembering . . . the story . . . the storyteller . . .*

> *Music out*

> *Slide: Crowd scene—Founding of L.A. commemoration, ca. 1935*

*Mestiza:* It was my mother who told me many of the stories of our family. The oral culture of the Californios has always been based in the family. She was repeating what she had been told or what she seemed to remember. But there was always some self-doubt. She used phrases such as, "You know, that's what they used to say." As if she was acknowledging the half-truth she was passing down to me. She was born in 1921 into a family of nine. It was her father's side that bore the Lugo line. She was raised in the era of the Hollywood version of Old California—rich, fiesta-making Spaniards, always dressed for a party, lace on the women, bolero vests on the men, dancing the night and day away. She was born into that myth. There was almost nothing left of that past, so why not begin to make it up. A little history here, a little Hollywood there. She always said that Don Antonio María Lugo received his land grant because he was . . .

*Music up*

*Mestiza:* "a very brave soldier and the King of Spain rewarded him with this great rancho." Of course, Antonio Lugo was not a soldier, and he probably never even met the King of Spain, but the majesty of such a statement made everyone feel real good. Spanish connections. Royal connections. The dominion. Maybe we're all Spanish to this day. Purebloods. Lighter, better, more European. She believes this, although the Lugos came from what is now Northern Mexico—Sonora—and had lived there for possibly two or three generations. And at the point that my family came to L.A. in 1771, Spanish citizens had been in this part of the world for more than 250 years. But she was never told this, and she never learned otherwise. I'm sure that if anything had truly been passed down through six generations, it was these stories about "Spanish" ancestors and the "attitude" that goes with them.

*Video up: Excerpt from* Ramona *film, ca. 1936, with Loretta Young and Don Ameche*

*Video out*

*Mestiza (dressed as Ramona Lubo):* They made me into a Mestiza señorita. I'm Ramona. I'm a Cahuilla Indian. Helen Hunt Jackson started it, in 1884. She wrote this book, *Ramona.* She was trying to make people aware of how hard life was for us, the Mission Indians of California. She was trying to change laws and give us back our land. We were supposed to receive a lot of land when the missions were closed in 1834. We were slaves to the Franciscan missionaries, peons to the Mexican rancheros. And the Americans coined the phrase "the only good Indian is a dead Indian."

Music sheet from the silent film Ramona, starring Delores Del Rio in the title role, 1928.

Helen Hunt Jackson wanted to create the Indian *Uncle Tom's Cabin.* She used my story. My husband, Juan, was shot and killed by a white man—Sam Temple. He thought my husband had stolen a horse. So he shot him in cold blood. One buck-shot. Twenty-two holes. My hand fit inside some of these holes. It was called "justifiable homicide." I was the only witness, but I wasn't allowed to testify because I was an Indian.

Helen Hunt Jackson got Sam Temple's character right. But not mine. She and all those moviemakers. They could have written my story, too, exactly as it was and made it very . . . sexy. . . . It was sexy for me, and romantic, and dramatic. I know she was trying to help, but the only thing she really helped was the turn-of-the-century land boom and the tourist trade. "Buy your dream home in the Land of Ramona." Ramona's every-where . . . Ramona the city, Ramona the song, the Ramona Pageant, Ramona Drinking Water, Ramona Mexican Food Products. My name's everywhere. No royalties. No residuals. No freedom. No civil rights. Nothing. That's all.

*Slides: Various of Ramona Lubo, ca. 1899*

*Mestiza:* I have a tale of my own. That our family was driven out of sixteenth-century Spain by the fires of the Inquisition. After seeing the books of their Jewish religion, philosophy, and poetry burn, they leave Europe and settle in Nueva España—a place of converts who do not get knocks on their doors in the middle of the night. Where the torture is far, far away. Where the sound of zealots is frozen. A place where everyone is "free" to call themselves Catholic. They might survive here if they just rearrange their rituals and learn to consciously forget while subconsciously retain the patterns of old languages, old rhythms.

*Music up.*

*Mestiza:* My tale may just be a subconscious remembering or a conscious longing to recognize my own complexity, my own mixed blood, and the need to figure out just what blood that is. The Mestiza—mixed blood—that I am, stares into the past and uncovers that which my own mother would never recognize—her own Indian. "Look," I tell her, "I can't deny what I feel, what I hear. Listen. There is a pulse inside me. I felt it once. It just came to me. A voice whispered, and I heard it. It had no name. It needs no name. But that voice that is me, spoke." And with those sounds, those words, I saw things around me I had never "seen" before. Trees spoke back to me. I looked up to them and said, "I will honor you for all my lifetime. I will listen to the voice in me. Your voice in me." This pulse, this voice is my Indian-ness. Once I heard it, I could never, ever forget. I had merged with the earth and nothing would ever be the same and it would never change.

*Music out. Mestiza takes Antonio Lugo painting off easel.*

*Mestiza:* In 1848, the Southwest becomes annexed. Attached to the U.S.A. like a heart that wants to reject the body it's been transplanted into. This land that is baby-sat by fifty-three families is to become divided, subdivided ad infinitum into bits and pieces of real estate that are now called by hundreds of names. Cities, streets, freeways, boulevards. Los Feliz. San Pedro. Boca de Santa Monica. San Bernardino. La Cienega. La Brea. Redondo Beach.

*Slide: Diseño (map) of Los Angeles area rancho, ca. 1840*

*Mestiza:* The ranchos are the blueprints. They didn't know this, but the Californios made designs that will forever remain embossed into this place. Our boulevards retain this memory. These boulevards—La Brea, Wilshire, Sunset—are the borders to the ranchos. Formerly footpaths, they recall their birth as they slash and burn through hundreds of neighborhoods. Forever fixed. With no fluctuation.

Mestiza refers to onstage diseño (mural painting)

*Mestiza:* Instead of surveys, the landscape of this place was drawn by the rancheros to provide the pictures of boundaries. Diseños, pictures with special points of interest—trees, hills, bends in the river. The rancho boundaries were not taken serious enough to be surveyed in the classic American way. Boundaries that say, "this is mine, exactly."

And without a survey, without any fixed boundaries, without any careful registry of who owned what, the United States created compelling legal situations—situations that forced all Californio grantees to sue the U.S. government—twice—or lose their land. While they were suing the government in San Francisco courts, with expensive attorneys, who had to hire interpreters, they were not allowed to have their property surveyed to prove which land was theirs. Nor were they allowed to sell their disputed property for cash needed to endure the lawsuits. The Californios had lived with a barter economy, but when the Americans arrived, cash became the order of the day. Eight years later, the relentless litigation has greeted these families like a middle of the night man with a gun nightmare. They can't prove where their land begins and ends. Their borders are not definite enough. So their land is given to squatters—anyone who has been living on and "improving" what was said to be vacant and undeveloped land. Land they said needed to be farmed on, not eaten up by cattle, for the soon-to-come midwestern masses.

*Slide: Western Avenue, ca. 1900*

*Slide: L.A. River flood, ca. 1861*

*Mestiza:* From November 1861 to May 1862, the rain did not stop. Fifty inches. The San Gabriel River jumped its banks. Only a residue of sand was left. Buried buildings, animals. The rain did not return for three years. Everything that was fallen and broken, even that which was left standing, baked. Should one pray for rain? Or just wait.

Waiting becomes death for thousands and thousands of animals and for a pueblo that will never look or be the same again. If you possessed anything before the flood, you barely still did, after the drought, when the rain begins again. The desert floor rises. Animals that have not been seen for hundreds of years suddenly reappear. They walk and talk as they have never done before. It seems to them that this could be their last chance to experience their home before the two-legged start pouring concrete. The snake with no fangs crawls up to survey what is left. Dead cows everywhere. Georgia O'Keefe would have a field day. It is not necessary to untwist the remains. Convoluted spines of metal, wood, and bone will untwist with time. The snake turns to herself and converses with her tail. The tail speaks: "What you see as unearthed will be built upon and upon and upon. But your serpent eyelids which never close will forever imprint this moment. You will not forget. Your serpent spine is connected to the most vital body signals that electrify all circuits so that nothing stops, nothing is forgotten."

*Video up of Mestiza as a contemporary real estate agent*

*Mestiza (on video):* Wouldn't you like to have a Moorish temple of your own? Five hundred expatriate Sephardic gardeners working for you in your own front yard. Inquisition closeout sales at your nearest minimall?

Well, with our Spanish Revival special, those memories are not lost. You're inside a lathe-and-plaster dwelling built with the latest European design technique. Such attention to detail. Simulated rafters, little Moorish domes, an oven big enough to roast a small pig, or at least some marinated chicken breast, with your own homemade salsa. Feel the exotic romantic history of Spain swirl around you. To dream the impossible dream . . . "Dream?" you say. This is L.A. After all, Los Angeles was a Spanish domain. I know, if you consult the census of 1810, only eighty persons out of 2,100 Latinos checked the español category. If you had light skin you could pass, so what the heck. After all those years in Mexico and a string of accidents, escapes, coital relationships, who could say? Indians, Sephardic Jews, Africans, Europeans, mestizos, mulattoes, Coyotes. But if we want to believe that a whole eighteenth-century soccer team of purebloods stopped over here on a one-way ticket to paradise, that's our American right and responsibility.

And, let me add, that when you do think of Spain, don't forget the Moors. A deeply spiritual group, but boy could they wield a sword. When you think of these Middle Eastern nomads, think of the few homes they built. Those homes would now be houses and apartment buildings for the American nomadic dream seeker. And they would probably look like this one. It's no coincidence that these authentic architectural details harken you back to another world. They will also harken home renters and home buyers looking for a Mediterranean sanctuary, a siesta getaway, or a new, fiery romance—great resale value. You will have a permanent no-vacancy sign, in tasteful neon, next to the security entry. Send your camel packing. Welcome to L.A.

*Video out*

*Slide: Contemporary Joe Camel cigarette ad in Spanish that reads "Un Tipo Suave"*

*Mestiza:* The loss of a language—this is what has been passed down. When you have been bought out, when you have to sell off what was once yours, when you see that loss in no uncertain terms, you figure that the culture you own, the language that you speak, is also for sale. The language of transients who have come to settle on your land, perhaps rob you of your home, becomes the language of power. You learn to speak English. You learn to honor English. Without it, you believe, you have no chance, no future. But the body still remembers. The tongue still rolls its *r*'s generation after generation. For my family, the loss of their language came early. Spanish is spoken. But it is a social language. A language for la familia. When it is "used," and I do mean "used," by those outside my culture, it is more as a language of control, a language of commands. "Compra, este." "Compra, ese." Buy this, buy that. Roll over. It is still not necessary to know this language. But that time is coming.

Two hundred years later, the language survives and is finding its way home.

*Slide: Parade, "Californio" on horse, ca. 1940*

*Mestiza:* I am trying to walk away from the mythical Spaniard in my mind. There are more than one. More than one version. But all figments of that imaginary figure. Who was that masked man? He still rides in parades, down boulevards, waving in complete silence. They see him but do not hear him. They distrust him, but they like his image—and her image as well. The fan, lace, layers of skirt. She is waving too. No, I think she is waving them off. Do not come here. Do not want this place. Do not marry my sisters. Do not become a converso. Do not make deals with my brothers. Do not learn my language, just to abandon it when another language replaces my tongue. Do not covet my culture for your own personal gain. Do not reinvent me on your salsa bottles, your wine labels, your tract home logos, your television sitcoms.

*Slide: Group of women, ca. 1890*

*Slide: Lugo grandmothers, ca. 1890*

*Mestiza:* I am looking for the actual woman who was here, my great-grandmother, and her mother, too. The mestizas. These women who built this pueblo, who began their days at 4 A.M., who bore far too many children, who buried their husbands twice their age. Maybe that's when they found their peace—when their children were grown and their husbands were gone but they were still only in their mid-thirties. Their brothers and their husbands were usually called españoles. It was more important that the men be thought of as sangre puros. They worshiped pure blood, thinking it kept them closer to Jesu Cristo. The culture could excuse the tainted bloodline in their women. If the men could just remember to stay in character—los hombres de razón—civilized Spanish men of reason. But she could live closer to the truth. I am looking for her.

*Slide: Antonia Lugo, ca. 1900*

*Mestiza:* She now rides like a phantom over this city looking for herself, looking for a place to touch down. The landscape recognizable, but there is no communication with the control tower. She needs permission to land, but they cannot give her clearance. They treat her like a foreigner, a UFO. They do not recognize her. She radios them that she knows the earth that lies beneath this tangle of cement surfaces. She spots the rancho, the path to the center of the old pueblo, el río de los temblores.

Actress Rose Portillo portrays the Mestiza Verrator in L.A. Real.

She knows she once lived here, before any of this. But without clearance, she may crash, she may disintegrate . . . again.

*Mestiza puts painting of Antonia Lugo on easel (based on photo, ca. 1900).*

*Mestiza:* She is part of that forgotten moment. A moment transformed in the hands of the new Americans. Erased and redrawn to satisfy the blank imaginations of those who came here to acquire their part of her. Remade in their own likeness with their own short memory. They replaced empty, vacant mission ghost towns with curio shops for the midwestern mind. And it may seem that this is what has been left for me. But if I reach back to trace myself, I feel something, I know something in my heart. I can't specify it. My genetic lines are crisscrossed. I start to go back, and leaps and bounds happen. The only thing I know for a fact is that it all happened right here in this city. I want to make a place for you to touch down.

*Music up*

*Mestiza:* Do not look for carretas or clipper ships patrolling the coastline like before. Follow the freeway. Please make an appearance. I'm calling for you. I'll wait for you to come.

*Music up and out*

*End*

# CHAPTER 16

# La Mona de T.J.

Gustavo Leclerc *(text)*
Julie Easton *(photograph)*

If la Mona could talk, she would surely speak in hybrid tongues, in combinations and mixtures of Spanish, English, Chinese, Zapotecan, Korean, German, and Mayan. Or perhaps she would communicate through complicated and twisted *albures*—those rhythmic sentences with double meaning, normally used to hide or disguise sexual connotations. It is considered the most down-trodden language within popular discourse; it is the mix of the lowly and the street: It is *rasquache* language.

If la Mona could walk, she would probably play at "jumping the fence" from one side to another; she would laugh at every-body from both sides. Or perhaps she would play hide-and-seek with *la migra* and *los coyotes*. At night, under the moonlight, she would bathe, and she would freely swim and scuba dive through the borderlines of the Pacific Ocean. She would crawl to sleep on the concrete bed of el río Tijuana under the bridge that crosses *la línea*.

La Mona awakes every morning looking north, yawning at the predictability of the other side's lifestyle. Everything seems monotonously *green:* the landfields, the uniforms, the INS trucks, the money, the imagination—even the future seems dimly green.

La Mona emerges from the guts of the colonia Nuevo Aeropuerto's urban collage. She leans on the skirts of the hills filled with self-made houses; she gets enveloped within the black blankets made from the intricate waving of Goodyear tires; she watches the routine departure of Boeings going north, *al otro lado.*

La Mona is like a public place turned inward, toward its own inner self. It is a place where interior become the most public spaces of the neighborhood's collective imagination, the space most densely imagined in its residents' memory.

La Mona has one owner but belongs to the whole colonia Nuevo Aeropuerto. She was built by one individual, but all the neighborhood's residents participated in its creation.

La Mona reacts irreverently to "high art" and spontaneously to the improvisations of the urban cultural landscape. She postures defiantly against "good taste," exuding an intense and creative self-confidence. Her aesthetic expression comes from the forgotten or abandoned, from the rejected or ruled out, from recycled fragments and left-overs, from urban everyday waste materials. La Mona is the border, La Frontera.

*La Mona,* by Julie Easton. Photo from *Huellas Fronterizas,* a continuing urban research project conducted by ADOBE LA.

# CHAPTER 17

# Waiting in Line

Anthony Hernández

Waiting in Line #21 (1997)
by Anthony Hernández
Dept. of Social Services,
Santa Ana office
Cibachrome, 30 inches by 30 inches
Courtesy of Anthony Hernández

Waiting in Line #33 (1996)
by Anthony Hernández
Dept. of Social Services,
East Los Angeles office
Cibachrome, 30 inches by 30 inches
Courtesy of Anthony Hernández

# CHAPTER 18

# Double Agent Sirvienta

Laura Alvarez

The Double Agent Sirvienta Rock Opera (1996-1998) explores the adventures of the Double Agent Sirvienta—an undercover agent posing as a maid on both sides of the border. The D.A.S. series includes a video, digital images, paintings, and other works on domestic surfaces. The intent is not only to elevate the position of the ethnic domestic worker with irony and humor but also to discuss technology as a symbol or tool of wealth, knowledge, and power. The narrative of the rock opera follows a young woman as she becomes a telenovela actress/pop star (every girl's dream) who always has to play maids, who, through disillusionment, an addiction to aspirin, and a vision, becomes an undercover spy for secret forces posing as a maid.

Synopsis

The following is a description of the conceptual and visual development of the D.A.S. Rock Opera as art performance:

*Pop Pura:* Suburban pop star eating Trix y tortillas when she gets home. She wears shiny TV star iridescent white cleaning uniform. She sees a vision of La Virgen (a projected slide) and adds roses to herself to come through her aspirin addiction.

*Cyber-Punkera:* Bad cooking show with Mexicana who doesn't know how to cook. It's messy. Cooking show becomes "give yourself a medical/cultural checkup examination." Tools are used. Medical show becomes constructing a big robot lady party. She's silver. Construction is easy.

*Nuevo Techno:* You see a border place sectioned off. After doing a dance around the robot, she falls into this area and gets into feather/plastic trouble (she's covered in it). Sirvienta gets chased around the whole place by a clown! Everything turns electric blue as people throw colored paper with a bright light from somewhere.

Opening
It's the Theme Song

She's trespassing
She can't be sassy
She's undoing
All of your plans

She's tiptoeing
So ignore her
And get ready
For her entrance

She's trespassing
Can't be sassy
She's undoing
All of your plans

A love story
Of her mission
Live in silence
Live in dreaming

Act Two: Construction:
Cyber-Punkera
Vámonos

Secretos
Tan negros
En limbo
Tan empty
Mis labios
Temblores
Las casas
Oscuras

¡Vámonos!
Con el servicio militar
¡Déjanos!
A combatir la plaga
Mírame señores
Mírame señoras
Tan chula, tan brava
¡Lo que tengo no es lindo!

Limpiando el sucio
Cortando zacate
Entrando a baños
Por puertas privadas

¡Encuentre!
¡Espere!
¡Lo que tengo no es fragile!

Sirpientes
De fotos
Que viven
Con susto
Se figen
En todo
Me tienen wachada

¡Vámonos!
Con el servicio militar
¡Déjanos!
A combatir la plaga

Mírame señores
Mírame señoras
Tan chula, tan brava
¡Lo que tengo no es lindo!

Act Three: Use: Nuevo Techno
Maquinita

Ay Diosito!
Ayudame con mi digital dilemma!

En tus suenos, Patrona!
It's too late now!

Éxito

I'm stealing
La technología
¿Zapa-quien on the interneta?
¿Zapa-quiere the interneta?
Hay te va la ha-ha-ha
Hay te va la ha-ha-ha

Machine . . . my machine
My maquinita
Tan bonita
My maquinita
Tan bonita

Get away
Don't stay back
Send the fax
It's alright
Message sent
And received

Running fast
Cushy shoes
Van outback
Run right past
Exit sign
Pointing sur

And across el río
Which way?
No se,
Es lo mismo
No es lo mismo

Quiero más misiones
Que no construyen más

In my dreams
I clean the scene
In my dreams
I tie the trash

I'm tasting all of the long
translations
Keeping up on instrucciones

misiones
Job done—flash back
Up to the billboard
Reside
On my T-shirt

Running fast . . .
Pointing south

New name
Another sound to answer
back to
New face
It comes clean every night I
wipe it
New style
A good look I can't get two
days twice
New day
A different way to sneak out
(whistle a while)

# 19

# Los paleteros

Camilo José Vergara

The ice-cream trucks of Los Angeles are too slow for the freeways. Built at least a quarter of a century ago and constantly in need of repair, each vehicle is unique. These trucks are painted vibrant colors that appeal to children, and the roofs are raised up so that the driver can walk around and work inside. Awkwardly tall and narrow, these ornately decorated trucks, known as *trocas,* are moving works of folk art. Their drivers are known as *paleteros,* a name derived from the *paletas* (wooden sticks) inside the ice-cream bars.

I have visited the special ice-cream truck parking lots. The drivers are mostly young Mexican men. A few married couples work together, and some bring their children along as helpers. One told me, "Behind each of these little trocas, there is a life, a family." A small business on wheels, the occupation has not changed much since the time of horse-drawn carriages. In Mexico, paleteros use pushcarts to do business, but recently, a few ice-cream trucks have been introduced by returning drivers who have learned the trade in the United States.

The paleteros often don't know the titles of the melodies they play endlessly as they cruise along. Knowing that Disney tunes are popular with their young clientele, however, one paletero may buy a cassette at a thrift store and share it with his fellows to copy. "If the trocas played Mexican *rancheras,* the children would not come out to buy; it is not the music they expect," said Luis. That explains why all the trucks play similar tunes.

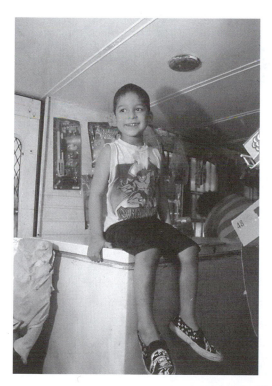

Working in the most dangerous areas of L.A., paleteros often confront armed robbers and vandals. A woman driver summed up the life on the road to me: "One gets stoned, shot, and ran over. Paleteros suffer very much." Her words sounded like the lyrics of a Mexican *corrido.* A driver who was listening in added, "People get inside the truck to steal the money and the merchandise."

In one parking lot alone, accommodating seventy trucks, three drivers have been killed during the past six years. A former driver named Marco Antonio recalled the panic he experienced when he stopped for a light and a man put a gun to his head and demanded his wallet. Soon after, he quit driving and took a job as a janitor in a hospital. Paleteros fear not only for themselves but also for their customers, especially the small children who might get run over.

To defend themselves, the owners transform their vehicles into fortresses. Selling takes place through a small window barely big enough for a can of soda. The truck exteriors display Disney characters, reinforced doors, and iron grates. José explained that he had to work behind bars because otherwise someone could pull him by the hair and steal his things.

Paleteros try to keep their vehicles free of graffiti. A driver told me that when the truck stops, five or six teenagers often surround it and spray paint it. "What can you do?" he asks. To keep his truck from being defaced, Mario painted a large image of the

Virgin of Guadalupe on its side, flanked by the flags of Mexico and the United States. Proud of his art, he tells me that the Virgin keeps his troca free of graffiti and protects him on his rounds.

A paletero makes between $30 and $60 a day, about as much as working in a sweatshop. But he is his own boss and spends time out in the open. This business requires little capital to get started, and the driver, according to the manager of a parking lot, "does not need to worry about a work permit." They need only pass inspection from the health department and obtain a $250 license.

A used truck costs from $1,500 to $2,500 plus the month's parking, the electricity needed to keep the refrigerator running, and the services of a night watchman—in all, an extra $50. Homeowners park their vehicles in their backyards, but most paleteros are renters, and their landlords won't let them connect the trucks to their houses.

A few paleteros supplement their earnings selling nachos, hot dogs, even produce, without a food license, thus risking fines of $100 or more. Others own more than one truck and rent them to others. Having a second job as a janitor, factory worker, or parking lot attendant is common. A few owners—tired of the dangers and meager rewards of the job—transform their vehicles into moving vans or use them for storage. There are even those who gut the interiors and rent the trucks out as a cheap place to live.

Some are able to save enough to start their own businesses. One paletero from South Central, now a baker, has parked his truck in his backyard, where it waits. If he fails as a baker, he can return to being a paletero without going through unemployment.

Despite working in such a dangerous and badly rewarded occupation, many paleteros seemed cheerful. Silda, a former paletera, agreed with me, commenting,

> They don't have anybody to give them orders, they don't have a boss. Nobody can tell them that they have to work tomorrow. Nobody tells them when it's time to quit. One suffers a lot during the cold and rainy season, from December to January. Then one does not even make enough to pay the rent. But during the hot days, one makes good money.

Luis, the manager of a parking lot, felt that paleteros were very happy with the freedom they have in their job and that they laughed and joked in the secure environment of the parking lot. Once in the streets, "They are worried. Sometimes, they are robbed as they start the day and have no money with them."

# CHAPTER 20

# Mi casa es su casa

Margaret Crawford
ADOBE LA *(photographs)*

**The Politics of Everyday Life in East L.A.**

Even the briefest encounter with East L.A. reveals a landscape of heroic bricolage, a triumph of what Michel de Certeau calls "making do."[1] These lived spaces, exuberant but overlooked, pose an alternative to the middle-class American house, actual or imagined. Taking control of ordinary personal and social spaces, residents have transformed a stock of modest single-family houses into a distinctive domestic landscape. Extending their presence beyond their property lines to the sidewalk and street, they construct community solidarity from the inside out, house by house, street by street. Through personal and cultural alterations to their houses, the residents of East L.A. reenact, in innumerable individual versions, the social drama of Mexican migration to Los Angeles, invoking memories that are both unique and collective. The house and yard are sites of ambiguous signification, revealing the complex tensions between culture and personality, memory and innovation, Chicano and Mexican, Mexico and America. Charged with human expression, these houses reestablished use and meaning as their primary definition. By investing their dwellings with the personal values contained in their interests, competence, and originality, the residents remove them from the context of mass-market values and, thereby, decommodify them.[2] Their pleasure in transformation and self-expression reclaim a central aspect of home ownership that many other Angelenos, obsessed with property values, have forgotten.

These activities appear to exemplify what de Certeau identifies as "tactics," the opportunistic maneuverings used by those without power. De Certeau's description of tactical operations emphasizes their momentary and circumstantial qualities, privileging fleeting experience over fixed and constructed space.

Comparing tactics to the meandering footsteps of thousands of ordinary pedestrians in the city, de Certeau celebrates their creativity and their meaninglessness. Because they evaporate in an infinity of personal trajectories, no order or pattern can ever emerge. Seen on the run, the quotidian transformations of East L.A. support de Certeau's conclusions. in contrast to the solidity projected by most suburban houses, these dwellings, shaped and given meaning by occupation and use, convey a provisional quality. Their occupants allow time and memory to control space.

Yet these apparently ephemeral uses and alterations, repeated countless times across several generations on adjacent lots and streets enclosed within the cultural and spatial borders of East L.A., come together, not only forming a distinctive pattern but suggesting a direction. Asserting more than just personal and social autonomy, they have acquired the potential for political significance. In its decommodified form, the house becomes a vehicle for mobilizing social identity, making a publicly legible statement that provides a new sense of agency. Thus, redefining the political field to include issues clustered around the home, daily life, and urban residential space offers the residents of East L.A., massively underrepresented in official political channels, new venues for collective activity. As a form of social action, their continuing use and transformation of existing houses question both the need for architectural intervention and the need for reinventing the house.

## The Border

Any discussion of East L.A. must begin across the Mexican border, 140 miles to the south, a line crossed by successive waves of immigrants drawn to the region's economic power. Mexican immigrants provided low-wage labor for the city's phenomenal industrial growth, but early in the century, Americans, intent on expanding the commercial district, building a civic center, and segregating their neighborhoods, pushed them across another border, the Los Angeles River. By 1930, the east side of the river was solidly Mexican. Currently, with more than 300,000 residents, the Eastside is the largest concentration of Mexicans and Mexican Americans in the United States, a Mexican city in the heart of Los Angeles. Beginning in the 1940s, thirty years of freeway construction imposed new borders. The incisions made by the San Bernardino and Santa Ana freeways on the north and south created barriers, not links, separating East L.A. even further from the rest of the city. Finally, the Long Beach and Pomona freeways carved an X through the community's heart, erasing it from the view of anyone passing through.

Today, for most Angelenos, the Eastside is both terra incognita and a zone of radical alterity. Its invisibility encourages out-

Border Crossing

siders to construct other, imaginary, borders. Is it dangerous or merely exotic—occupied by drug dealers, cholo gangs, or anonymous hordes of illegal aliens? Newspaper and television reports further demonize the Eastside, providing additional ammunition for the police and sheriff's departments that control the area like occupying armies. In fact, the Eastside's crime rates are no higher than those on the Westside. When the Eastside is seen from within, along narrow streets lined with working-class bungalows, the myths of East L.A.'s mean streets melt away. Here, unnoticed by the city and hidden from professional architectural culture, its residents have created a new hybrid form of dwelling. Negotiating between the circumstances of life in Los Angeles and the customs, rituals, and traditions brought from Mexico, they define a new border condition, no longer a line of exclusion but a cultural free trade zone, accessible to continuous movement back and forth.

**The Fence**

The fence is the initial gesture that defines East L.A.'s domestic landscape, a method of staking claim to the barrio for as long as Mexicans have lived here.[3] Chain link is common and acceptable, but residents prefer elaborate constructions of block and wrought iron. Refusing the amorphous and impersonal suburban front lawn, they use the fence to delineate the front yard as an enclosure. This moves the domain of the house forward to the street, extending its domestic space to the corners of the lot. A permeable border, protective but inviting interaction, the fence enables each family to define its environment while maintaining contact with the activities of the sidewalk and the street. In East L.A., every street presents a characteristic topography of fences; some are patrolled by dogs, others are hung with homemade signs advertising nopales or discount diapers, and others support brightly colored brooms for sale. As innovation has encouraged imitation, fence styles have become increasingly complex, spurring the rapidly evolving craft of wrought iron, one of East L.A.'s largest homegrown industries.

**The *Yarda***

Fully occupied, the enclosed yard encapsulates the functions of the plaza, courtyard, front yard, and street. It is simultaneously an arena of sociability, a site of control, an outdoor work area, and a stage for symbolic elaboration. Both public and private, the *yarda* welcomes engagement with neighbors and passersby but can also shelter intimate discussions and family celebrations. Its nuanced space structures social encounters: Strangers are met at the gate, friends invited onto the porch. By providing a vantage point of the street, it allows residents to supervise adjacent territory. Owners and renters customize their yards by occupation and design. Busy with mundane tasks or artfully arranging plants or statuary, they expose their daily lives and deeply personal preferences to anyone passing by. Infinitely flexible and always in flux, the *yarda* can accommodate special events, from a garage sale to a *quinceañera*. It can become a lush jungle of plants or, paved over, a playground or car repair shop. A driveway can

Model

serve as a dance floor, an outdoor hallway, or a space to display goods for sale.

Alterations  "Mexicanization" is a recognizable yet versatile idiom, a mutable aesthetic adaptable to many conditions. Immediate changes are easily accomplished with yellow, mango, or peach paint; strings of Christmas lights; and attention to intricate detail. Other alterations require more time and money: replacing wood surfaces with textured stucco or columns with ornamental wrought iron or stuccoed arches. Still others derive from family needs: Garages are remodeled as rental units, extra bedrooms, or small businesses.

In addition, porches are often added, expanded, or covered, furnished with tables, chairs, or couches and decorated with wrought iron, paint, or potted plants. Whether the product of the owner's weekend labor or that of unlicensed neighborhood contractors, most alterations are constructed without working drawings, building codes, or permits.

Plants

Apart from the ubiquity of rose bushes—collective remembrances, perhaps, of the roses miraculously transformed into an image of the Virgen de Guadalupe—gardens in East L.A. are primarily lovingly tended personal statements. As visions of paradise, they demonstrate the many varieties of metamorphosis, blurring the distinctions between agriculture and ornament, artificial and natural, sacred and mundane. Neat rows of corn and nopales appear among the standard Southern California cypresses, fruit trees, and bougainvillea—traces, along with illegally kept chickens, of a rural past. Eno, a variety of Spanish moss used to decorate Christmas crèches, hangs from tree branches. On any street, one might find container gardens in recycled washtubs set on concrete slabs; complex compositions of flowers and paving, with fountain and birdbath centerpieces; and comic assemblages of plants, plaster yard ornaments, found objects, toys, and solemnly pretty shrines elaborated around statues of Christ or the Virgin Mary, all vying for attention.

Model

The House

The interior of the house is the center of the family's power. Because the *yarda* is the primary site of social life, casual guests rarely enter the house. Although large extended families live in these small spaces, they rarely alter the arrangement of rooms. Instead, parents, grandmothers, and children crowd in together. With children sharing tiny bedrooms and everyone sharing the single bathroom, privacy is rare, but not particularly valued. Living rooms overflow with plastic-covered furniture; photographs of birthdays, graduations, and weddings; and carefully ordered collections of "beautiful little things," religious mementos, and family souvenirs. The television is here, imposing its schedule on the room: Spanish-language soap operas from 4 to 7 p.m., followed by American situation comedies, and, on weekends, Dodger games and soccer matches. Still, the heart of family life remains the kitchen, the only room that might be enlarged. As the focus of cooking, eating, and household chores, this is the territory of the

Installation with Model

mother and the female members of the family. Fathers, working outside the house all day, appear in the evening to watch TV, water the plants, or putter in the yard.[4]

The activities that take place in these houses and yards unfold through time, producing the distinctive rhythms of East L.A. On weekends, the enclosed household routines spill out into the yards' extroverted space. On Saturday, the tempo intensifies, as men gather to work on cars, friends and relatives drop by, tables and chairs are set out for parties and barbecues, vendors offer their wares, and teenagers cruise by in minitrucks or low riders. By Sunday, the pace slows. Families, dressed up, leave for church, neighbors chat over the fence, and children play in the yard. Family time is measured in even slower cycles: Renters move to other houses, relatives arrive from Mexico, children grow up, and a new generation moves east to suburbs in the San Gabriel Valley or San Bernardino County. Yet throughout these changes, threads of memory persist, to be rewoven into surprising new combinations. The polyglot sounds overheard from car stereos and boom boxes on Spanish-language KLAX (Southern California's most listened-to radio station) and such popular shows as *Alma del Barrio* or *Voices of Aztlán* demonstrate this process of transformation. Segueing from norteño ballads to rhythm and blues oldies, banda to Led Zeppelin, they mix old and new, English and Spanish, nostalgia and rebellion, in complex combinations that like East L.A. itself, defy interpretation.

Installation (Detail)

### The Politics of Everyday Space

The cultural and political weight of these continuous transformations renders speculative housing development, architectural intervention, and traditional real estate practices almost completely irrelevant in East L.A. Because poor and working-class Mexicans and Mexican Americans rarely purchase new houses, market-oriented designers and builders do not take their tastes or preferences into account when selecting spas or improving master bedroom suites for new subdivisions. In this world of hand-me-down housing, the professional services and specialized culture of architects are even more remote. When architects choose to donate their services, their results have often been inappropriate or demeaning. Contemporary architectural styles, whether postmodern or abstract formalist, hold little interest for a culture already rich in visual imagery, expression, and meaning.

Ironically, these identifiably "Mexican" qualities have contributed to low prices and stability in East L.A.'s housing market. Almost entirely Mexican, it is an unlikely target for gentrification. In other parts of the city, however, the quotidian house transformations tolerated in East L.A. bar-

Installation and Studio

rios are under attack. Class and ethnic struggles over space disguise themselves as struggles over architectural values. Preservationists in Pasadena, horrified by "Mexicanized" Craftsman bungalows, have issued Spanish-language pamphlets that attempt to convince homeowners not to alter their houses. In response to their arguments for the social and financial rewards of maintaining a house's original historical and cultural character, East L.A.'s lived politics of the everyday poses the questions: whose history, whose culture, whose house, whose space?

# Notes

This project was generated by an invitation to participate in House Rules, an exhibition organized by Marc Robbins for the Wexner Center for the Arts at Ohio State University in the fall of 1994. This text was published in *assemblage* 24. We (Margaret Crawford and ADOBE LA) were one of ten writer-architect teams given the opportunity to rethink the American house in light of recent social and theoretical changes affecting concepts of home, family, and identity. Our project, conceived as a critical and polemical statement, should be understood as an intervention into this highly specific context, rather than as a general statement about or a scholarly analysis of Mexican American houses in East L.A. By presenting a compendium of features taken from existing houses in East L.A., we hoped to challenge the idea, implicit in the organization of the exhibit, that architects are (or could be) privileged actors in creating the domestic settings in which most Americans live. Professional designers are even less relevant in influencing low-income and working-class housing patterns. Instead, we wanted to demonstrate that creativity and competence as well as cultural and personal expression are not the sole property of trained professionals but are widespread in neighborhoods such as East L.A. Our selection of this particular neighborhood was also intended to underline the necessity of discussing houses in their actual settings. Far from generic, our house is located in a real place shaped by a unique set of historical, social, political, and economic events and circumstances. Finally, in contrast to the highly abstract forms of writing and representation that dominate architectural discourse, we presented our project using words, photographs, and models that could be easily understood. We felt that clarity and communication were particularly important because the project would appear in *assemblage,* an academic journal specializing in fashionably arcane architectural criticism. Our emphasis on the positive aspects of self-building should not be mistaken for a celebration of the status quo. We are arguing, rather, that identity and social change are always produced by life and not by design.

1. Michel de Certeau, *The Practice of Everyday Life* (Berkeley: University of California Press, 1984), 29-42.

2. See David Harvey, "Accumulation Through Urbanization," *Antipode* 19, no. 3 (1987): 269-71. James Holston describes the process that Harvey outlines in his work on Brazilian self-building, "Autoconstruction in Working-Class Brazil," *Cultural Anthropology* 6, no. 4 (1991): 447-65.

3. This discussion is indebted to James Rojas's pioneering work on East L.A., "The Enacted Environment: The Creation of `Place' by Mexicans and Mexican-Americans in East Los Angeles" (master's thesis, Department of Architecture, Massachusetts Institute of Technology, 1991).

4. Ibid., 80-82.

# CHAPTER 21

# St. Francis of Aztlán

Rita González *(photographs)*
Ramón García *(text)*

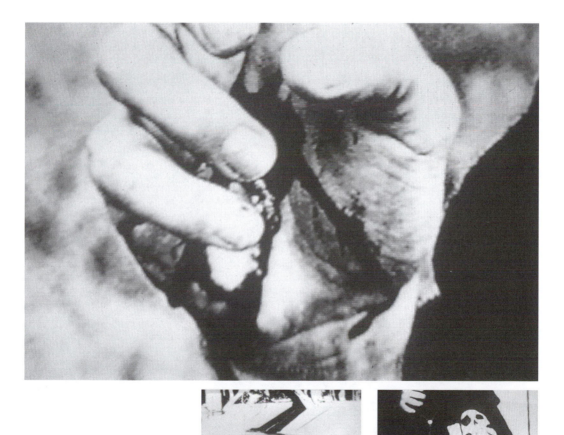

St. Francis of Aztlán

Little Francis was a beautiful cha-cha boy.
He was handsome and superficial.
He was an Echo Park party boy.

A plague was ravishing the city.
Death was gathering momentum in the shadows.

And one day the Lord decided to break his usual silence
and he spoke to Francis,
"Go find the city of the angels
the necropolis in your flesh," he said.

And Francis went in search of the vast inner landscape of his
    history.
He walked, brother Death a constant companion.
The empty paths of lost memories opened up before him,
    guiding his way.
Everywhere he encountered the debris of his past—
repression, desolation, murder, betrayal, oppression,
    abandonment.
He had been forgotten.

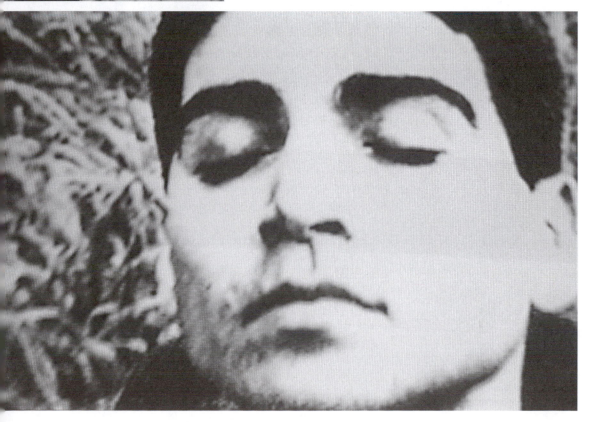

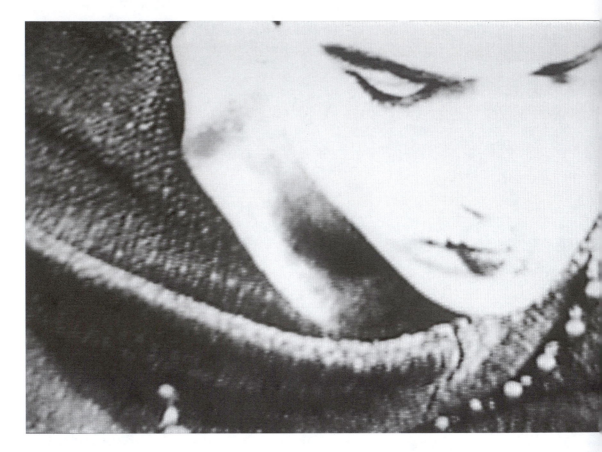

Wandering was the only way he had of remembering.
Remembering was the ladder to God.

The spirit of dead angels followed him.
They were the messengers of massacres.

He blessed the dead and forgotten birds that crossed his path.
It pained him to think of the sweet melodies which they,
in the company of other fluttering friends, once sung.

The carcasses of animals too, he blessed and said a prayer for.
The side of the road is the suburb of the freeway, it is a zoo of
   the dead.
He prayed as he walked.
It has been forgotten that walking is the same thing as praying
when one walks in the midst of death.

One day when Francis was wandering in Aztlán, the city of birds,
a small community in the eastern outskirts of the city,
he had a vision:

an angel appeared crucified to a cross
and Francis felt the presence of the Lord inside him.
Memory momentarily ravished him.

The past inhabited his flesh.
Christ was a city on fire;
the fires raged in his rupturing flesh.
The blood of Christ, the love of Francis became one
    conflagration.
He felt the passion of Christ in a riot of angels' wings in his
    heart.

And the crucified angel spoke in the voice of God,
"Stay in Aztlán and rebuild yourself, work from your wounds,
reassemble your mutilated wings and the silenced chorus of
    your brotherhood.
Love yourself and love your neighbor, hate injustice.
Teach them the sight of blood, the evidence of history.
To live is your salvation.
To love your only sustenace.
To fight your only hope."

And St. Francis continued walking.

# CHAPTER 22

# The Latino Use of Urban Space in East Los Angeles

James Rojas

The Mexicano residents of East L.A. use their front yards and streets to create a sense of community. The identity of place is created not only by the physical forms but by the way they use exterior space around their homes and businesses. The exterior space provides a background for people to manipulate as they please and to act in, much like a movie set. To understand this living architecture or enacted environment, one must examine people as users and creators of place through their behavior patterns that are affected by culture and geography. Thus, architecture not solely defines the environment but contributes to an "enacted environment" as one element.

Latino residents seem to spring forth from the asphalt. They can be found on streets, corners, sidewalks, and front yards as well as in marginal places such as parking lots and alleys. Street vendors are popular and flow in and out of commercial and residential areas and attract crowds wherever they go.

A typical house in the barrio resembles any other house in Los Angeles. The tremendous difference is in the appearance of community that is formed by the residents' use of space around the home. By selling, working, playing, and hanging out in these outdoor

AUTHOR'S NOTE: An earlier version of this essay appeared in *Places,* spring 1993.

spaces, their presence creates a spontaneous, dynamic, and animated urban landscape that is unlike any other in Los Angeles. The streets, front yards, driveways, and other spaces around the homes bring residents together, whereas in other L.A. communities, these same spaces isolate residents.

Props    Movable objects or props, such as tables and chairs, allow the Latino residents to control the outdoor space by giving them flexibility and freedom over their environment. Props can be moved between inside and outside space, as well as allowing for permanent or temporary "personalization" in public space. These items create a sense of security in a place by acting as markers for territory. Like furniture in a room, props in the street connect the user to the open urban space.

A pushcart selling ice cream captures a fleeting moment of social exchange between eager children. A sofa under a tree or on a porch can allow the resident to wallow away the afternoon, whereas a barbecue pit can generate some revenue and neighborhood gossip.

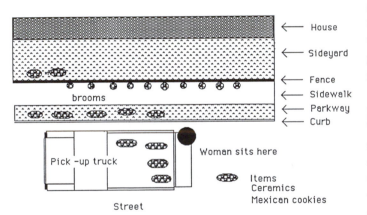

In the barrios of East L.A., props are symbols of place. They can be seen, heard, felt, and smelled. On weekends, one can follow the crowds, smell the roasting ears of corn, taste the tangy fresh-cut pineapple and chile powder, touch the smallest trinkets, see the colors and the people, hear the laughter, dance to the music, and sway to the rhythm of the barrio.

From costumed mariachis who walk from bar to bar or restaurant to restaurant singing songs for dollars and to the blaring car stereos of young adults playing banda (Mexican country music) or disco, music adds a rich, intense ambience to the suburban landscape of East L.A.

Some gas stations have been converted into taco stands by the heavy use of props, with only minor changes to the structure. Instead of saying "Shell," the thirty-foot signs will advertise the taco stand. Wrought iron sheds are sometimes added in an attempt to enclose some of the open space. Pumps are replaced by a formal arrangement of tables and chairs. This formal arrangement is as arrogant as any European outdoor café. People who sit and eat here have direct visual access to the street, thus reenforcing street activity. This function is expressed with a minimal amount of retrofitting by using props and people to create the atmosphere.

Just as the machine-age abandoned gas station have become integrated to reflect Latino cultural values, so has the automobile.

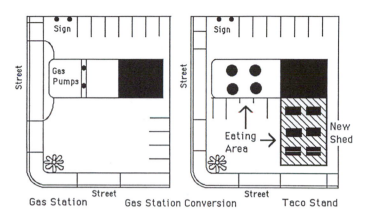

Gas Station    Gas Station Conversion    Taco Stand

**Low Riders**  Latinos use the automobile for social interaction by cruising down streets, parks, and other areas where young people congregate. In Latin America, this social mingling takes place in the plazas. The absence of plazas in Los Angeles forces Latinos to drive the car, instead of making the traditional paseo (walk).

Aside from cruising, Latino males, like their white counterparts, are fascinated with automobiles. These Latinos are passionate about their hobby and spend much money and time with their buddies from car clubs on the streets and in garages customizing cars. The classic bombs (1940s and 1950s cars) and sedans and Chevy low riders of the 1950s, 1960s, and 1970s have given way to smaller cars and minitrucks and a new breed of low rider.

The chrome rims and simple paraphernalia that created low riders of past generations have been replaced with elaborate details such as hand-painted murals, gold-plated hydraulic scissor lifts, suicide doors, posche rims, and other high-tech innovations. Low riders are no longer the ambition just of teenagers but more often of adult Latinos in their twenties with more discretionary income. The cost of customizing a car today can cost thousands of dollars. Cruising is still popular among teenagers, but many of today's embellished low riders rarely see the asphalt or are driven only at night when there is no traffic because one little pebble can shatter an expensive custom paint job.

Many of these low riders are seen at monthly car shows throughout the southland, which cater to a largely young Latino crowd. These customized cars are displayed with all the reverence and attention to detail one finds in a Latino church altar.

Trophies, teddy bears, champagne glasses, Mexican flags, and other items enhance the low riders. Chrome engines, pink velvet interiors, water-filled doors, and other elaborate elements illustrate the tactile genius and imagination of the Latino car owners.

Crowds walk through the numerous rows of cars in every color, make, and model—taking pictures and talking to the

car owners. As if at a modern-day rodeo, the crowds gather to watch low riders hop, or dance, as it is called, with hydraulic lifts to record-breaking heights. The latest craze, bed dancing, is created by hydraulic scissors that raise the truck bed more than ten feet at different angles.

Latinos do not stop at generic symbols of auto conformity, like their white counter-

parts, but create personal self-expression of urban art. The low rider represents the Latino ability to appropriate and redefine American standards.

**No Blank Walls**  Few walls are left untouched in East L.A. With graffiti, store signs, and murals, wall space becomes a cultural expression of many forms for cholos (gang members), political groups, and shop owners. Garage doors, fences, sidewalks, building walls, benches, buses, and, recently, the freeways signs have become targets of personal expression. All this expression creates a new reality of visual stimulation, as well as filling in the urban environment. The order of the place and the purity of the blankness are violated by these images.

**Murals**  Murals are a form of political, religious, and whimsical advertisement in barrios, each expressing different values. Murals of Our Lady of Guadalupe are popular because she is the patron saint of Mexico. Many of the murals from the 1970s express social ideologies such as Chicano power. Most common, however, are the whimsical murals that the shop owners commission for advertisement.

Murals make marginal urban spaces tolerable and can be appreciated both from the car and by foot. Murals offer residents an inexpensive, quick way to personalize space. Most are painted on the large expansive walls on the sides of buildings. These spaces are usually targets for graffiti. The murals may not stop it, but they prevent the graffiti from dominating the space. In addition, they provide advertising for the owner and employ local youths to create the "masterpieces." Murals wrap the commercial activity into an otherwise forgotten space by livening up the area. These corner areas are important spots because street vendors hang out there. The aesthetics created by these murals and graphics are more important than the building, thus becoming the architecture.

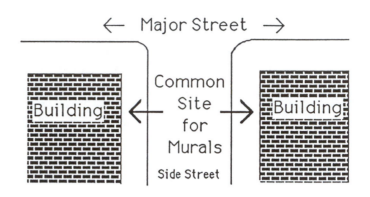

La Yarda: A Personal Expression

Nowhere else in the urban landscape of East L.A. is the Mexican use of space so illuminated and celebrated than in the enclosed front yard. As Mexican immigrants settle into their new homes, the front yard became a personal expression. Depending on the needs of the owners, the use and design of the front yard will vary from elaborate courtyards reminiscent of Mexico to junkyards. The Mexican brings a new interpretation of the American front yard. The front yards of the barrio reflect Mexican cultural values applied to the American suburban form.

Fences: A Social Catalyst

Fences can be found enclosing many front yards across America. For Latino residents, fences enclosing the front yard have a different meaning. Most Americans regard a fence enclosing a front yard as hostile or exclusionary from the rest of the neighborhood: a barrier against the world. In the barrios, fences are catalysts that bring neighbors and pedestrians together for social interaction.

Edges, borders, and boundaries are dynamic places where people come together. The fences break down the social and physical barriers by creating an edge for residents to lean on and congregate. The psychological barrier that the front lawn creates in American suburban homes no longer exists in the barrio.

In Spanish, there is a saying that "through respect, there can be peace." Respect for the individuality of each resident is reenforced in the use of the fences that clearly delineate property ownership. Therefore, the residents can personalize their front yards without physically interfering with the neighbors.

The use of fences in the front yard modifies the approach to the home and moves the threshold from the front door to the front gate.

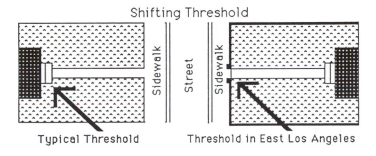

Shifting Threshold

Typical Threshold          Threshold in East Los Angeles

The enclosed front yard defines a space between the public and private spaces of the home in relation to the street. The enclosed front yard acts as a large foyer and becomes an active part of the household. One develops a sense of entry into the home by stepping into the front yard through the front gate off the sidewalk. For residents and pedestrians, it is perfectly acceptable to have conversations at the front gate and not be invited into the home.

### East L.A. Vernacular

The homes in East L.A. were built by non-Latinos; they have evolved into a vernacular form, however, because the residents have made changes to suit their needs. The houses are customized and personalized, with every change, no matter how small, having meaning and purpose. East L.A. vernacular represents the struggles, triumphs, and everyday habits and beliefs of working-class Latinos. The vernacular form offers cultural, economic, and regional solutions to the residents' criteria. The beauty of the vernacular can be measured not by any architectural standard but by life experiences, which are ambiguous. The vernacular represents people's manipulation and adaptation of their environment.

One must understand the differences and similarities in the Mexican and Anglo values toward home and the urban landscape to be able to comprehend and appreciate the unique combination. A bastard of two architectural vocabularies, Latino homes and barrios create a new

language that uses syntax from both Mexico and America. The diagram below illustrates this:

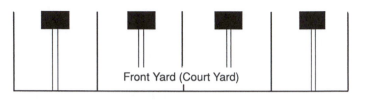

## Street (Plaza)

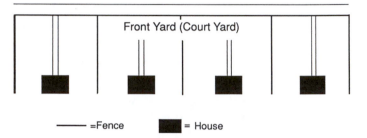

Latinos bring a new perspective to the American suburb in the Southwest through their appropriation and redesign of the environment. The new perspective fuses together Latin American social values imposed on American urban form and can offer solutions to rethinking the urban form. This fusion illustrates the evolution of the suburb that was designed to break down ethnic cultural differences.

CHAPTER **23**

# Fotonovela

Harry Gamboa Jr.

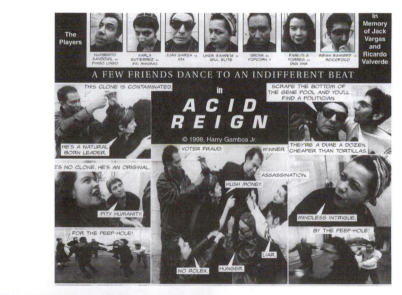

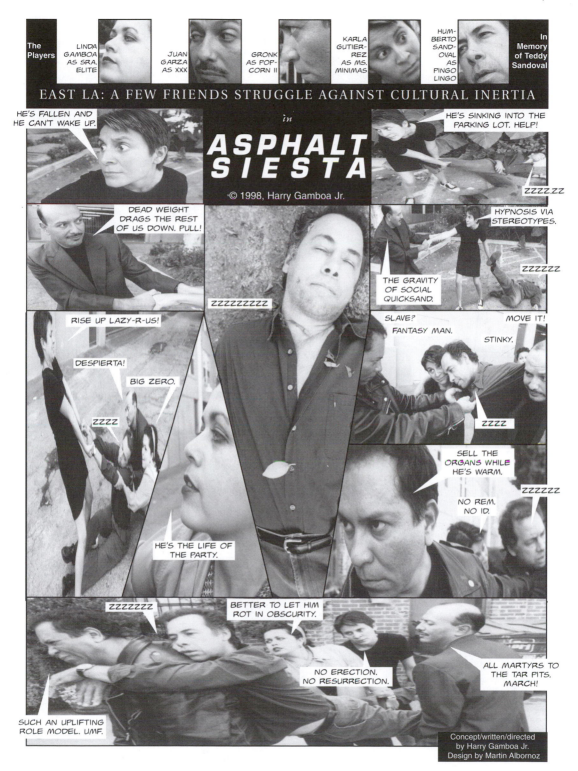

# CHAPTER 24

# The Underground Music Scene of Los(t) Angeles
# Choosing Chicano in the 1990s

Yvette C. Doss

June 28, 1997, Art and
Commerce Gallery,
Downtown L.A.

It's a warm summer night, and a few hundred people are gathered in a nondescript building on an otherwise isolated street in L.A.'s downtown warehouse district. They've come here for a record release party for a compilation of Chicano music called *Sociedad = Suciedad* ("Society = Filth") on Big Daddy Records/ B.Y.O. Records. It features songs by local bands Aztlán Underground, Ollin, Blues Experiment, Quinto Sol, and Ozomatli recorded from October 1996 through May 1997.

Inside, long ponytails and braids are the style for many of the young men and women in the audience. There is an abundance of indigenous jewelry, traditional Latin American *guayabera* summer shirts, Aztec calendar T-shirts, and embroidered peasant blouses.

A closer look will also reveal dreadlocks and shaved heads mixed in with the ponytails, as well as nose rings, tattoos, stuffed-animal backpacks, black leather jackets, baseball caps, and basketball jerseys.

Excitement crackles in the air as the bands come on, rapping and singing with energy and purpose.

Steam rises toward the ceiling as a friendly slam pit develops, then doubles in size. Bare-chested, twentysomething men skank and mosh to a ska-funk-punk beat.

■

Although musically, there is very little that binds many of the rock, hip-hop, punk, blues, ska, and folk bands that make up the core of L.A.'s underground Chicano music scene—aside from the occasional bilingual lyric and traditional Latin American or in-

AUTHOR'S NOTE: This chapter first appeared in *frontera Magazine* #6 and #7, 1997.

143

digenous musical inflection—philosophically and conceptually they are united.

Song themes vary from the dangers of urban life to environmental pollution, from racial unity to unfulfilling sex, from indigenous genocide to love. One song by Aztlán Underground touches on the theme of the rootless Chicano, which Mexican author Octavio Paz called a "lost soul" in his book *Labyrinth of Solitude*.

"I know I'm not free because I'm a lost soul. . . . I don't know where to turn, I roam out of control," Aztlán Underground front man Yaotl Orozco sings.

The scene—made up of about two dozen bands including Ozomatli, Quetzal, Aztlán Underground, Lysa Flores, Cactus Flower, Ollin, Quinto Sol, Blues Experiment, JABOM, and Announcing Predictions, among others—is the product of a generation of fans and artists whose cultural and musical sensibilities haven't been expressed by either the Latin America-oriented rock en español or traditional "Anglo" rock scenes, nor by the popular Mexican sounds of banda or mariachi.

The artists produce music that is an urban blend of traditional and modern, retro and futuristic: turntable scratching and conga beats, *jarochos* alternating with rock and soul, and manic, punk-tinged ska and blues accompanied by tales of frustration and heartbreak in a city some call "Los(t) Angeles."

It is music that is a direct reflection of the participants' hybrid identity in Los Angeles society.

"We may be wearing *guayaberas*, but we're also wearing Doc Martens and nose rings," says twenty-five-year-old Flavio

Morales, host and cofounder of the independent Chicano TV show *ILLEGAL interns.* "The fact that we were born here makes our use of the imagery, sounds, and symbols of Mexico 100 percent American."

Morales should know. He and the rest of his crew have been documenting the Los Angeles Chicano/Latino music scene since 1990 through their hour-long East L.A. television program devoted to contemporary Chicano music and culture.

There have been other bands in this progressive tradition, notably, the Plugz and Los Illegals in the late 1970s and early 1980s, which were playing in clubs in East L.A. and Hollywood when Morales was still a *mocoso* in diapers. But few recognizably Latino musicians—aside from Los Lobos—have been able to survive the record industry's marketing machinery, which many musicians say attempts to fit groups into preexisting slots such as "Latin" or "alternative" but which has never been able to make sense of a group that can be categorized as both "Latin" and "alternative."

This discouraging history has not been lost on the leaders of this new Chicano underground. Many of them feel that most of their talented Chicano predecessors had to compromise their music and identity to conform to record company stereotypes about Latinos and their viability in the mainstream marketplace.

Not wanting to make what they feel would be futile compromises with their music, these new bands operate, for the most part, outside the parameters of the recording industry. They play gigs at places such as downtown art galleries, the old city jailhouse north of downtown, and restaurants and cafés in Little Tokyo and produce their own records from money scrounged from fund-raisers. It's a living, although not a very glamorous one.

The bands—many of whom explore lives of "urban exile" in their lyrics—appeal to hip, young Latinos who appreciate the

subtleties of the Chicano cultural subtext in their music, a sensibility shaped by negatives: neither this nor that, always on the border. "These bands speak to the soul of our Chicano community," says Consuelo Flores, a local writer and performance artist who has been following the scene for years. "They have an intelligent and relevant message that transcends race and gender boundaries. It's a message of perseverance, community involvement, inclusion, and hope."

The irony is that the scene—best defined by its grassroots support and independence from the industry rat race—is gaining such momentum that record companies are starting to take notice. Ozomatli, a bilingual Latin jazz, salsa, and hip-hop group that regularly draws up to 1,000 fans to its shows around town, has already been signed to Almo Sounds, the label started by A & M founders Jerry Moss and Herb Alpert. Lysa Flores, one of the stronger female vocalists, signed on with Geffen as musical director of the soundtrack for the 1997 movie *Star Maps*, which featured her track "Beg, Borrow and Steal."

■

**June 15, 1997,
The Roxy, Hollywood**

Lysa Flores is on stage. Her long, curly brown hair shimmers, its burgundy streaks catching the stage lights. The twenty-three-year-old Mexican American singer performs her brand of quirky feminist rock with a guitar hoisted on her hip. The audience has come out to a fund-raiser for the Chicano music compilation *Aztlán Alternativo: Barrios Artistas,* with cuts by Lysa Flores, Cactus Flower, Announcing Predictions, Motita, Quinto Sol, and others. The album, along with *Sociedad = Suciedad,* is one of the two most important documents of the L.A. Chicano music scene to come out in years. The evening's showcase, one of a handful of fund-raising shows organized by compilation producer and musician Eddie Ayala, is called "Club Aztlán 2000." Soon, Ayala will change the name of the showcase performances.

A scheduled band called Rice and Beans won't be going on tonight. One of the members jumps on stage and begins shouting at Ayala. The musician snatches the microphone out of Ayala's hand and explains his beef to the audience members. He says Ayala owes him money and that he promised his band a better show time.

Later, Ayala will rename his music showcases "Boxing Chicano," a title that will appear in the *Los Angeles Times* concert listings page.

"Chicanos are always fighting," Ayala will explain, with a laugh. "We like drama."

At that night's show, *copal* smoke rises to the ceiling as Aztec dancers, Ayala's invited guests, circle the dance floor during an intermission in shiny metallic loincloths, elaborate headdresses, and feathers.

■

Chicano. The musicians, and fans, are choosing the moniker *Chicano*, a word that has often been dismissed by more assimilated Mexican Americans as archaic, distasteful.

"Don't call me Chicana," a twenty-four-year-old painter tells me. "I don't like the word." Although she was raised in East L.A., the title conjures up unpleasant connotations for her. She rejects it.

But the Chicano that the L.A. underground music scene participants use to describe themselves is not *that* Chicano.

It's not the Chicano of the low rider-cruising, teardrop tattoo-bearing, oldies-loving urban warriors of the 1960s and 1970s. Not the loud-talking, big hair, raccoon-eyed, comb in back

pocket-carrying Chicanas with the *pinto* boyfriends who made their Mexican mothers wring their hands with worry.

It's Chicano as in younger cousin of the vinyl pants-wearing, spiked hair, 1980s and 1990s New Wave music-loving nouveau Chicanos.

It's Chicano as in pierced nose and tongue, *guayabera*-wearing, postmodern cultural samplers of the mid- to late 1990s Los Angeles.

Members of this underground scene are Virtual Chicanos (independent filmmaker Jim Mendiola's phrase of choice for this chameleon-like generation) who can shrug off disparate cloaks of identity at will, wearing a different one every day of the week.

They are Chicano as in urban borderland dweller. They are a new breed of modern culture shifters who, approaching the turn of the millennium, share a sensibility shaped by a love of alternative rock, hip-hop, and *ranchera* music, Thai food and *tamales*, Japanese animation and "Chavo Del Ocho," shopping trips to the Beverly Center and El Mercadito, with consistent voting records and fierce activist streaks, as well as, in the case of the scene's musicians, a facility with electric guitars and *vihuelas* and a penchant for lyrics about Henry Miller and indigenous rights.

"Our generation has the best opportunity so far to take the culture further as far as aesthetics go," explains Quetzal's guitar player Gabriel Tenorio.

And they have chosen to call themselves Chicano, salvaging the term from its previous, hyperpolitical essentialist confines and letting it loose on the modern world to evolve and take on new, broader permutations.

■

It's Thursday evening, and five of the members of Blues Experiment are sitting in the Bell Gardens home of lead guitarist

Robert Tovar and his girlfriend Lilia Ramírez, in a living room cluttered with an assortment of instruments, including a standing conga drum, two guitars, a bass guitar.

Joshua Durón is lying stomach-down on the floor, tapping out a tune on a little handheld electric keyboard. The band, with two members squeezed into a love seat, breaks into a straight blues song, and lead singer Gus Sabina starts singing from his perch on a stool, softly at first, then building to a steady, soulful high.

A couple of songs later, the band breaks into an impromptu jam session. Everything in sight becomes an instrument; one musician jangles a set of keys, the back of a guitar becomes a makeshift drum for Sabina, and Tovar picks up a pot from the kitchen and starts banging on it with a spoon.

Anything-goes jam sessions such as the one at Tovar's house are a habit for Blues Experiment and other bands such as Ozomatli that trace their roots to a one-year experiment in urban community building at a site members called the Peace and Justice Center, which was born from a political strike against the Los Angeles Conservation Corps.

A group of mostly twentysomething Chicanos, former employees of the Conservation Corps, took the building over in March 1995 in protest over low wages and the firing of a popular director. The government ended up granting them the rights to the downtown building, so they founded their own community center there, a pseudocommune experiment that was a life-forming experience for many of the young participants.

Bands such as Loto Azul, Ollin, Marble, Rice and Beans, Black Eyed Peas, and earlier incarnations of Blues Experiment and Ozomatli (which called itself Todos Somos Ramona, in reference to the Zapatista leader) performed at weekend concerts at the site. During the week, the center's program directors held workshops in community organizing, poetry writing, underground theater production, and studio mixing.

Politics is integral to the music of many of the Chicano scene's bands, particularly groups such as Aztlán Underground and the reggae group Quinto Sol. It's the engine that fuels their music and helps them make sense of life in a city that many feel refuses to include them in its definition of itself.

Yaotl Orozco, thirty-year-old vocalist and lyricist for Aztlán Underground, describes a Southern California upbringing full of self-doubt. He says he felt that he didn't belong in this country.

"Growing up in the San Fernando Valley, which was mixed white and Chicano, I'd hear things like, 'Stupid wetback' all the time," Orozco says. "So I always felt that I was in someone else's country and I shouldn't be here. I'd even get mad at my parents for speaking Spanish."

For Orozco and his band mates, music is an outlet for anger and frustration, as well as a way of coming to terms with past injustices and finding hope for a better future.

"Aztlán Underground is about using music as a medium to talk about our untold history and our experiences here in the U.S.," he says.

"We all share a sense of community, and the reality of being nonwhite and living in the U.S.," says Bull Dog, the half-Irish, half-Mexican vocalist and lyricist for Aztlán Underground. "We've finally got some place where we can go that's our own."

The compilation *Sociedad = Suciedad* was a direct reaction to the political climate surrounding Proposition 187, threats against affirmative action that took concrete form in the University of California bans, and, later, Proposition 209, as well as an overall feeling of impending doom many young Latinos in Los Angeles were feeling in the mid-1990s.

"We were all playing political benefits during the uprising in Chiapas, and when Proposition 187 passed, we decided to put the compilation out to record a period of history," said Robert López, guitarist for the group Ollin and co-owner of Luna Sol Café, which regularly hosts performances by local Chicano poets and musicians.

But not all are as tied to the idea of music as a means of expressing political ideology. For Lysa Flores, who because of her less obvious political inclinations seems to be on the periphery of the scene, just existing as a Chicana is enough of a political statement.

"*Chicano Alternativo* is simply music coming from a Chicano perspective," Flores says, explaining the term that Ayala coined and chose as the title of his compilation. "What we're doing is political because it's not mainstream. We're not allowed in the mainstream yet. Every song we sing is political because we're trying to break down stereotypes."

■

For nearly fifty years, Candela's Guitars on Avenida Cesar Chavez in East L.A. has sold handcrafted classical guitars to some of the area's, and the world's, most accomplished musicians. Faded, yellowing photographs of Jos, Feliciano, Charo, Mariachi Sol de Mexico, Pedro Infante, Los Lobos, and others hang on the shop wall, a testament to the pageant of great musicians who have come out of, or made their way through, East L.A. since Porfirio Delgado and his son Candelario opened the shop in its current spot in 1948.

These days, many of the newest crop of Chicano and Latino musicians get the tools of their trade there. But you won't find Gibson electrics or mass-produced Fenders at Candela's. The

store, which Manuel and Tomás Delgado took over after the recent death of their father, Candelario, crafts custom-made classical guitars, smaller pot-bellied *vijuelas,* and the rarer *jaranas* and *guitarras de golpe* that are staples of traditional genres of Latin American music such as the Mexican *corrido, jarocho, bolero,* and *huapango.* Strewn behind the counter on a recent afternoon were dozens of string instruments in rich hues of Honduran mahogany, Brazilian rosewood, ebony, German spruce, and *palo escrito,* or Mexican rosewood.

Manuel Delgado says the sale of traditional instruments such as the *jarana* has been on an upswing for a few years now.

"Everybody's going back to the music their father or grandfather was listening to," says Delgado. "Being Latino wasn't cool, so for a long while, people pushed away from it. I think this generation has wised up."

Although it may be true that incorporating traditional musical forms into modern punk, pop, and rock is a viable form of "cool" these days, as evidenced by the rise to prominence of culture sampler Beck Hansen (who said in an interview with *frontera Magazine* that he considers himself an honorary Chicano), most musicians in the scene will tell you that their choice to take up traditional instruments has more to do with a rejection of the sanitized, whitewashed modern rock proliferating the radio airwaves than with an attempt to follow a trend.

"I listen to the radio to know what not to play," laughs Quetzal Flores, guitarist for the band Quetzal. He says picking up the *jarana* changed his life by giving him new musical direction.

For many of the twentysomething musicians in the Chicano underground, playing traditional indigenous and Mexican instruments is a way of finding their roots and reconnecting with a mestizo or indigenous past from which they feel cut off.

It's also symbolic of the concerns many of the musicians share, which include community development and cultural exchanges with other indigenous groups, such as a recent "Encuentro Cultural," or cultural encounter in Chiapas.

The desire to reconnect with their indigenous roots shows up in the names the bands have chosen for themselves: Quetzal is a Nahuatl word meaning "precious feather," as well as the word for bird; Ozomatli is the Aztec god of dance; Aztlán is the mythical homeland of the Mexicano people.

Martha González, twenty-five-year-old singer and conga player for the group Quetzal, explains it like this:

There's an ancient pueblo verse that says "*El pueblo que pierde su memoria pierde su destino,*" or, "If you lose your memory, you lose your destiny." You need to have a past in order to have a future. It's very important to be responsible for who you are and where you came from.

"Using the instruments and the music that your family left behind now that you're in America is as radical as you can get," says Juan Carlos Whyte, coproducer of the compilation *Sociedad = Suciedad.*

■

With his current band, Cactus Flower, East L.A. veteran singer, producer, and promoter Eddie Ayala sings a song called "No Chicanos on MTV," a direct homage to Lalo Guerrero and his humorous tune, "No Chicanos on TV."

Many of the musicians in the L.A. scene said they would add a new twist to it, calling it "No Chicanos on KROQ," referring to the local commercial modern rock radio station.

Although Los Angeles County is about 40 percent Latino, and Arbitron reports that 28 percent of KROQ's listeners are Latino, the L.A. station has yet to hire any Latino on-air personalities and plays few, if any, of the local Chicano rock or alternative bands. In the past year, few of the Chicano bands have made it on the local music show, called "Music from Your Own Backyard With Zeke."

According to Mark Torres, host and producer of a Latin rock radio show on KPFK (90.7 FM) called "Travel Tips for Aztlán," it's not because of a dearth of talented Chicano bands.

"With two hours a week on my radio show, I'm not going to change the complexion of the Los Angeles music scene," says Torres, who reports that his show currently receives about fifty demo tapes from unsigned Latino groups, as well as groups on independent labels, during any given month. "But if a popular, youth-oriented commercial radio station like KROQ added these important Chicano bands into regular rotation, it could make a big difference."

"Right now, it's not like we're doing anything with that," said Lisa Warden, music director of KROQ, when asked if the station has plans to play any new alternative Chicano rock bands. "We recognize that there's a market out there, and we might do something in the future."

The solution to the barriers Chicano musicians face on the radio and in the marketplace, many believe, would be a Chicano/Latino record label that would emerge as the community's own Motown, able to nurture and develop talented Chicano musicians for audiences both within and outside the community itself.

The idea of such a label, which could turn Latino musicians into household names, has been tossed around for decades, ever since Del-Fi Records' Bob Keane produced Ritchie Valens's smash

hit, "La Bamba," in the late 1950s, and Eddie Davis produced music by many of the "Eastside sound" bands of the 1960s such as the Premiers and the Blendells on Rampart Records.

"Crossover dreams," they used to call it.

But few of today's musicians seem concerned with crossing over to a commercial audience anymore. "I don't want to be a way-paver," says the Blues Experiment's Tovar. His comment is in reference to the role many of Motown's musicians, such as the Supremes and Smokey Robinson, played during the label's heyday, when musicians were sent to "etiquette schools" that taught the performers how to be just a little less "black." They paved the way for later musicians.

"If that's what it'll take, then I'll just pack up my guitar right now," says Tovar.

■

Shawn Stern, founder of B.Y.O. Records and lead singer and guitarist of L.A. punk band Youth Brigade, says he decided to release the compilation *Sociedad = Suciedad* on his offshoot label, Big Daddy Records, because of both the power of the music and the potency of the message.

"The bands from East L.A. who were singing in English and Spanish were being completely ignored by the big labels," says Stern, referring to both the major Latin labels and mainstream English-language labels. "I liked the spirit of the music, and the independent attitude reminded me of the punk rock scene that I came from."

"I figured, why don't some of these Chicano bands record their own music instead of waiting for a record label to come around?" coproducer Juan Carlos Whyte said.

To save money, the album was recorded in one of the coproducer's apartments, and the whole thing was done on a shoe-string budget. All of which is indicative of one thing. More and more these days, the attitude of Chicano musicians is, "Why deal with an industry that just doesn't get it?"

The women in the scene, particularly, say the industry just doesn't get it. They are tired of worrying about whether they might have the right look to make it big in a rock world that still worships at the altar of the blonde.

Two standout talents in the scene are Lysa Flores and Quetzal's lead singer Martha González. Although playing distinct styles of music—Flores's English-language folkish rock is reminiscent of that of Natalie Merchant and P. J. Harvey, while González sings soulful, gut-wrenchingly powerful songs in English and more traditionally oriented tunes in Spanish—both women are tough, smart, and no-nonsense feminists who tackle subjects from the Zapatista struggle to abortion with equal bravado.

González—who as a child appeared on the short-lived TV sitcom *A.K.A. Pablo* and in the early 1990s was signed to a record label as part of a hip-hop group, Son Estilo—comes from an industry background already. "That feeling you get when they're trying to find a look for you is wrong," she says. "I felt like I had to have plastic surgery to have my cheeks shrunken and my nose made narrower."

Flores, who has recently found inroads to the industry with her work on the Geffen soundtrack, said her past encounters with record label executives have been at times infuriating, at times amusing.

"Latinas are always being stereotyped," she said. "I sit in record label offices where they go 'Oh, you're Latina, do you sing in Spanish?' Or, 'Can you do the whole Selena thing?' and I just say, 'Uh, it was nice meeting you.' "

It's incentive enough for all the independent producing going on at the grassroots level.

■

L.A.'s Chicano music community has always depended on grassroots support for local talent.

Producers such as Rampart Records' Eddie Davis in the 1960s and Fatima Records' Yolanda Ferrer and Richard Duardo in the 1980s—who coproduced albums by The Brat and The Plugz—invested in Chicano bands by founding small, independent record labels.

But since the 1983 release of *Los Angelinos: The Eastside Renaissance,* a seminal compilation made up of 1970s- and 1980s-era Chicano rock bands (produced by Rubén Guevara on Zyanya/Rhino), as well as the January 1997 release of Zyanya/Rhino's anthology *¡Ay Califas! Raza Rock of the '70s and '80s,* also produced by Guevara, there has been a huge gap in the documentation of music recorded after the 1980s.

Some are now working to fill in that gap.

Aside from the compilation *Sociedad = Suciedad,* there has been a lot of independent production activity in the recording studios around Los Angeles.

There are Eddie Ayala's compilation on his own indie label, Lista Negra/Mexnut Enterprise; upcoming releases by Quetzal (their first) and Aztlán Underground (their second); and, on the periphery of the Chicano scene, a hard-core Chicano punk compilation called *Propaganda: East Los Underground,* put out by Rod Larios in 1997 on his indie label, F.O. Records.

It's a direct manifestation of the Chicano do-it-yourself, *rasquache* spirit. "We're saying we don't need to fit in," says Quetzal's Gabriel Tenorio. "We'll make our own space."

Whether the record industry—or "Hollywood" as many call it—ever comes knocking on their door, most of the musicians in the scene say they'll continue playing because for them, making music is about more than just selling albums, or even creative expression. More often than not, it's about cultural survival.

"It all comes back to dignity," says Martha González. "It's about what comes out of your heart, without having to mold it to fit into some square in Hollywood."

# CHAPTER 25

Cultura, migración, y desmadre en
ambos lados del Río Bravo
# Más allá de las mamonerías

Rubén Martínez

Evangelio
Del Libro de los Hechos de
los Apóstoles:

*Se les aparecieron unas lenguas como de fuego, las que, separándose, se fueron posando sobre cada uno de ellos, y quedaron llenos del Espíritu Santo y se pusieron hablar idiomas distintos.*

*Culturas Posfronterizas*

Un cholo de la Meseta Purépecha de Michoacán deambula por la calle principal de Nuahuatzen, abriéndose paso entre abuelas de rebozo y campesinos de botas enlodadas. Porta su cachucha de los Oakland Raiders al revés y tiene la cabeza rapada al estilo de *East L.A.* Anda con sus Nikes y sus «baggy pants.» Trae una camiseta sin mangas que deja ver su tatuaje de las máscaras de la comedia y la trajedia con el lema «la vida loca» clavado en el hombro.

Entra a un antro de videojuegos con sus cuates, donde se la pasa matando ninjas negros y árabes. Cada vez que mata a uno de los malos, exclama: «¡En la madre, *motherfucker!*»

Después, se sube a su *ranfla,* un Datsun destartalado modelo 79 con placas de Carolina del Norte . . . y por su pueblo se echa a *cruisear,* cantando el estribillo de una *oldie* . . . «*My angel baby, my angel baby/ooohy I love you, yes I do* . . . »

Ya sonando las campanas de la iglesia, a las ocho de la noche, regresa a casa donde su abuelita en trencitas lo espera. Lo saluda en tarasco, y el cholazo posfronterizo, con mucho respeto, le responde en su dialecto ancestral. Se sientan en la sala, prenden el televisor Samsung, conectado a la parabólica en el techo, y se clavan un par de horas wachando MTV, el noticiero de CNN, y la novela *De pura sangre.* . . .

*Meanwhile, Back in Los United States:*

Conozco a un joven chicano cuyos jefes emigraron de la mismísima Meseta Purépecha hace veinte años, agarrando jale en la pizca de lechuga en Watsonville, California, en la pizca de sandía en Kentucky, en la pizca de tabaco en Carolina del Norte, en la pizca de naranjas en Flórida, chambeando un rato en el ferrocarril en Nebraska, como camareros en un hotel de Dallas, y, por fin, viviendo en el sur de California, donde sus padres arreglan los papeles y compran una modesta casita en el Valle de San Fernando (donde, hace tres generaciones, los mexicanos pizcaban naranjas).

Este joven se destacó como estudiante excepcional en la *high school,* le encantaba la biología, y ahora cursa segundo año de licenciatura en la UCLA. Habla perfectamente el inglés y el español y hasta algunas palabras en tarasco. Fue fanático del «death metal» y «trash,» pero hoy es miembro de MEChA, Movimiento Estudiantil Chicano de Aztlán, y todos los fines de semana se sumerge en el bosque nacional de los Padres, una zona montañosa al norte de Los Angeles. Ahí, un viejo indio de la tribu chumash enseña a los chicanos inquietos las tradiciones indígenas y profetiza sobre una guerra espiritual en la que la raza de bronce habrá de recuperar su dignidad. . . .

El chicanísimo posroquero purépecha regresa a casa después del ritual en el temazcal y se clava un par de horas con sus jefes y hermanitos a ver un poco de MTV, las noticias de CNN, y la telenovela *De pura sangre.* . . .

Flujo, Ajetreo, Continuidad

*Palabra del Libro de las Aventuras de la Gaby (escandalosamente suprimido por el Cardenal Ratzinger), el trasvesti tapatío más cachondo del club El Plaza, un antro de pocamadre ubicado sobre la avenida La Brea a la altura de la calle Tercera en Hollywood, California:*

*Evangelio*

Mi amor
estamos siempre partiendo

partiéndonos en dos
desmadrándonos partiendo;
es un parto-partir-partiendo siempre
para arribar en ninguna y en todas partes
¡ay! pero que chulo estás, mi querido . . .

Si vemos el presente con el lente del pasado mamón, diríamos que *la identidad nacional* está siendo atacada una vez más por el invasor yanqui *free trader* y que cada parabólica es una amenaza directa al reinado de la santísima Virgen de Guadalupe.

Si vemos el presente con el lente del pasado mamón, diríamos que los chicanos son unos pinches pochos sin ningún derecho a llamarse mexicanos, y que los narcocholos de Michoacán atentan contra el espíritu nacionalista de nuestro México lindo y querido.

Si vemos el presente con el lente del pasado mamón, diríamos: «qué lamentable que los purépechas vean MTV, las noticias de CNN, y la telenovela *De pura sangre* en vez de estar cultivando su maicito, descalzos, sin herramientas modernas.»

A los que persisten en pensar que hay una frontera lineal que separa lo que es ser mexicano, indígena, mestizo, chicano, etcétera, la historia ya los rebasó. Los que persisten en la noción de «la vida onírica del indígena» niegan el presente del indígena: niegan el hecho de que puede ser—y es—tan moderno como los «posmodernos» oriundos de cualquier gran urbe del planeta. De hecho, más indios hoy en día viven en la ciudad que en el campo; un chingo de indios viven al otro lado; es decir, los indios que el mestizo admira congelados en los dioramas del Museo de Antropología e Historia son más inquietos, más roladores, y más conocedores de la modernidad y sus bases socioeconómicas reales que el mestizo mismo.

El indígena es el que chambea al otro lado y regresa cargando un televisor y una videocasetera nuevos para disfrutar de las películas de Steven Seagal.

Más que pérdida de identidad, vemos la continuación del proceso de mestizaje, en donde el indígena—y el chicano—tienen voluntad propia para armar el paquete cultural a su antojo.

Igual que los mestizos lamentan la supuesta pérdida del pasado indígena, ven con tristeza a los chicanos y a su supuesta «crisis de identidad.» Pero quienes ven en los chicanos una «pérdida de mexicanidad» no se conocen a sí mismos. El chicano en muchos sentidos es más «mexicano» que el chilango de clase media, cuya mirada encandilada siempre apunta hacia Nueva York y París.

Así se crea una falsa dinámica: el mestizo clasemediero de la capirucha siente que el futuro está en el norte (en Estados Unidos o en Europa) y que el pasado está en la Meseta Purépecha (o en la selva Lacandona o en la sierra Tarahumara). La verdad es que el tiempo y el espacio ya no conocen este tipo de fronteras primitivas. El futuro está en ambos lados, el pasado también, el presente en todos: la parabólica y el cholo en Michoacán, el neoindígena y los equipos de pelota mixteca en California.

Todo se mueve, todo cambia, todo permanece, y, al parecer, los únicos que se sienten cómodos dentro de este ajetreado paisaje son los indígenas y los chicanos, quienes reconocen que el futuro y el pasado co-existen en el presente.

El joven mixteco que vive en Fresno, California, y que ya no habla su dialecto, sigue siendo mixteco precisamente porque la cultura es un organismo que para mantenerse vivo debe adaptarse a su nuevo entorno, seguir creciendo. Pero, como anotó el filósofo Oswald Spengler, el entorno también se va adaptando al nuevo organismo que se hace presente: los gringos hoy en día consumen más salsa ranchera que *ketchup,* sólo para citar un dato gastronómico superficial.

El futuro no necesariamente aniquila al pasado: en el presente pueden convivir tradición y modernidad. En los pueblos de la Meseta Purépecha, la casa con la parabólica apuntando hacia los cielos puede estar habitada por una bruja que trata las «males enfermedades» con hierbas y el tarot, o bien por un *teenager* Trilingüe—español, inglés, y tarasco—y a quien le encanta el grupo de rock Transmetal y las pirecuas a la vez.

Ver este proceso como nocivo para la salud cultural es proyectar la imagen del indio como víctima pasiva de la historia. Ese es precisamente el más grande de los estereotipos construidos por los mestizos sobre la identidad indígena.

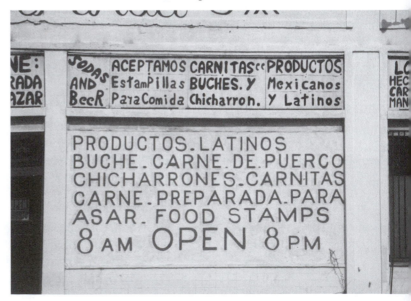

Hace unos meses llegó a la Ciudad de México una joven activista norteamericana cuyos padres habían emigrado a los *United* desde la India. Cargaba unos de esos *backpacks* asquerosos que suelen traer los gringos o los europeos cuando viajan por el tercer mundo (como si emprendieran un *safari* en pos de elefantes y aborígenes). Le espantaba la capital: «*So many white people,*» decía. Tanto bullicio, tanta luz, tantos edificios, tantos coches. De plano abandonó la ciudad para encontrarse con los tzotziles en Chiapas. Ellos no necesitan de electricidad, de televisores ni de zapatos o libros, decía emocionadísima. Los indígenas viven *au naturel.* ¡Qué *cool!*

De la misma manera, los mestizos de la capital necesitan de sus mitos indígenas para sentirse modernos, ya que padecen de un complejo de inferioridad ante el gringo o el europeo. La máxima hipocresía aflora en el momento en que el mestizo capitalino se vuelve nacionalista y adopta una posición neoindígena ante los extranjeros.

Cuando vine por primera vez al D.F. como adulto, hace más de diez años, los profes universitarios y los izquierdosos en general adoptaron una actitud paternalista hacia mí. Pobre chicano, me decían. En tu país padeces el mal del racismo. Aquí en México, no tenemos crises de identidad. ¡No mamen!

Los chicanos (o en mi caso, los chicanos-salvadoreños nacidos en Los Angeles que ahora viven en el D.F.) sabemos que la estabilidad, podríamos decir un tanto como los budistas, es un estado de movimiento. Simplemente, hoy día, los que no se mueven se mueren. Lo cual es todo lo contrario del lema del nuevo operativo de la *Border Patrol:* «*Stay out, Stay alive*» (colgando retóricamente en el cerco fronterizo a los cadáveres de los ahogados en el Río Bravo y los muertos de sed en el desierto para que sirvan de escarmiento). Pero son muchos los mexicanos que saben que *to stay alive is to move.* Económicamente, culturalmente, lingüísticamente, sexualmente. Por lo tanto, en vista de lo anteriormente afirmado, presentamos la

*Plataforma del Partido Mojado*

- El problema no es el idioma que hablamos ni con qué acento lo hablamos. *El problema aquí es la* Border Patrol.
- El problema ne es ser gay, *straight*, o bicicleta, o travesti. *El problema es el Sida.*
- El problema no es si somos católicos o pentecostales o sufis. *El problema es la falta de tolerancia, es la persistencia del Estado y de la Iglesia Católica y de otros poderes sociales y económicos en seguir fomentando la intolerancia al promover una falsa homogeneidad nacional.*
- El problema no es la venta ambulante o la prostitución o la drogadicción.

*El problema es el neoliberalismo que deja a muchos al margen de la posibilidad de participar plenamente, en lo económico y lo cultural, del proceso de globalización del cual gozan las clases medias estadumidenses y europeas, a quienes tanto les gusta bailar salsa, comer comida tailandesa, y asistir a los* performances *de Guillermo Gómez-Peña.*

Utopía y Apocalipsis

*Evangelio*

Del libro de *La Licuadora* (también escandalosamente suprimido por el cardenal Ratzinger), la más chingona de los *coyotes* del pueblo de Cherán, Michoacán:

Ya nos chingaron una vez
los de la migra mierda
gringa.
Pero para la próxima
les advierto que contamos
con artos cuernos de chivo.
No es por nada
que me dicen la Licuadora . . .

También en Estados Unidos, se promueven falsedades homogenizadoras, provenientes tanto del *establishment* conservador y liberal (republicanos y demócratas) como por la izquierda marginal (que aglutina los particularismos étnicos, sexuales, y de clase).

Se ha dicho, por ejemplo, que al alcanzar poblaciones mayoritarias latinas en varias ciudades de Estados Unidos la raza podrá ejercer, finalmente, algún poder político para poder contra restar medidas xenofóbicas como la Proposición 187 de California.

Pero los latinos de los *United* no somos homogéneos en nada. Somos salvadoreños y guatemaltecos, cubanos y puertoriqueños, hondureños y colombianos y nicaragüenses—y entre los mexicanos tenemos que distinguir entre los recién llegados, los chicanos, y los hispanos de Nuevo México, cuyas raíces en el suroeste son añejísimas. Además, somos de clase media y trabajadora, somos rubios y morenos e indígenas, somos católicos y pentecostales y judíos.

Somos todo lo que somos al otro lado (es decir, en Latinoamerica).

Dificilmente podemos pensar que los cubanos de Miami siempre estarán de acuerdo con los chicanos de California, o que los migrantes zacatecanos se llevarán siempre bien con los de Michoacán (recuérdense las riñas entre estos dos grupos en Saint Louis, Missouri, donde el saldo del conflicto fue de varias decenas de muertos y heridos).

Estamos todos—en ambos lados del Río Bravo—inmersos en un proceso de mestizaje aceleradísimo: culturas y subculturas brotan como las mil flores de Mao.

Este proceso nos crea nuevas utopías y apocalipsis a la vez. Por ejemplo, en el barrio de Compton, al sur de los Angeles, cuya fama mundial se debe a las violentísimas pandillas afroamericanas y a los *rapperos* como Ice Cube y Niggaz Wit' Attitude (N.W.A.), la población latina—en su mayoría migrantes recién llegados de México y Centroamérica—amenaza con desplazar a la comunidad negra.

Mientras este cambio demográfico se lleva a cabo, dos realidades opuestas se enfrentan en las calles de Compton. Por una parte, un conflicto de índole racial y de clase entre negros y latinos. «Pinches mayates,» dicen los mexicanos de los negros. «*Fuckin' wetbacks*,» dicen los negros de los mexicanos.

Pero hace dos años, en la *high school* de Compton, un joven salvadoreño fue elegido como presidente del consejo estudiantil. Ganó votos de negros y latinos. Porque el cuate habla inglés y español. Porque escucha rap y *oldies* y boleros y rock. Porque su novia es negra. Porque prácticamente nació en el barrio (llegó a los seis años de su país natal) y maneja tanto el inglés en su versión afroamericana como el español.

Tenemos dos presentes, dos futuros contrarios: la torre de Babel o un nuevo Pentecostés.

Lo que nos amenaza con la incomprensión de una nueva Babel es la ruptura económica que hace que grupos «marginales» compitan entre sí por las migajas del nuevo orden económico, que claramente no ofrece el sueño americano a las mayorías.

A medida que se frustran más y más los sueños de una vida mejor de los mexicanos en Nueva York, los afroamericanos en Chicago, los turcos en Francia, los nigerianos en Inglaterra, y los purépechas en Michoacán, aumenta la desesperación y las medidas desesperadas para sobrevivir.

Cruzar la frontera en Arizona con el riesgo de morir deshidratado en el desierto . . . Entrarle al narco-tráfico, a la prostitución, a la venta ambulante, a las mil maneras de participar del mercado negro . . .

O desquitarse por medio de la violencia entre semejantes—los zacatecanos que se dan en la madre con los michoacanos en Saint

Louis; la pandilla *18 Street* (mexicana) con la pandilla *Mara Salvatrucha* (salvadoreña) en el barrio de Pico-Union en Los Angeles . . .

La unidad política entre los latinos, si es que se da, será puramente coyuntural.

La lucha contra la 187 en California fue una coyuntura clásica: en 1994, días antes de las elecciones en que se aprobó la medida anti-imigrante, se realizó una marcha de más de 100,000 personas en Los Angeles, con plena participación de chicanos y salvadoreños, los recién llegados y los de tercera generación.

Sin embargo, dicho movimiento se desbandó después de la derrota electoral. La desesperación y la frustración pueden crear una fuerza aglutinadora, pero también una que acelere la fragmentación social.

Ahora estamos más fragmentados que nunca, lo cual es jodidísimo, lo cual es bellísimo. Al resquebrajarse justamente los falsos esquemas homogenizadores del pasado crece tanto la conciencia de nuestra diversidad (y, ojalá, la tolerancia) como una especie de angustia existencial. Si el *México profundo* no existe, ¿con qué llenamos el vacío? Si el *melting pot* no existe, ¿cómo se reconstruye el *American Dream*?

El momento no es para desenterrar mamonerías (los neomarxistas del misterioso grupo guerrillero EPR en México; los chicanos neonacionalistas remasticando los mitos aztlanecos en Estados Unidos) ni para andar con la cabeza agachada. Es para seguir expandiendo nuestro concepto de identidad, de tolerancia, de democracia.

Y lo crucial: es para buscar la manera de conectar lo que son los procesos de migración cultural y social con la problemática económica. Porque todos sabemos a estas alturas, como se ha dicho en Chiapas, que donde hay hambre no puede haber democracia. O, como diría cualquiera de los chavos purépechas posfronterizos: cuando no hay chamba, ¡fuímonos p'al otro lado!

# CHAPTER 26

# Seeking Oblivion in Los Angeles

Reynaldo Rivera *(photographs)*
Ramona Ortega *(text)*

Los Angeles is a city fragmented by freeways and morally encrypted billboards, a polycentric place with streets that take you from one isolation to the next. Yet people are drawn to Los Angeles from all directions, seeking an illusion of reality that was embedded in them by TV, magazines, and pop culture. Often, their adventures leave them strung out, poor, and without the capability to remember what brought them here.

Despite its debilitating effects, Los Angeles remains the destiny of overly ambitious movie stars from the east and fleeing Mexicans from the south—because the city of angels inspires angelic dreams of rags to riches, from castoff to stardom—but with all its luster, it can't quite fulfill either one.

In the shadow of time between Hollywood's splendor of the forties and the suburban sprawl of the fifties, Los Angeles turned into L.A. It became too big for its own good, making promises it couldn't keep. The gigantic Hollywood sign that greets newcomers became a reminder that those who live in the hills are large and white and very removed from the valley below.

There was a time in the seventies that Hollywood and L.A. attempted to come together and reflect an image more like itself. It was cool to be poor; even rich kids wanted to be poor. *Chico and the Man, What's Happening,* and *Sanford and Son* were not *un-Politically Correct*—just correct. But somewhere in the cross fire between trickle-down and the nineties, the man in Hollywood decided we needed a new image, one that would take us far away from Propositions 187 and 209 and minimum wage. But L.A. didn't notice. People were too busy making crack an epidemic, crowding

jails, or catching SIDA in a bathroom stall to realize that after oblivion, there's only reality.

In a place where dreams are portrayed as real and reality is obtained by drugs and crime, we find ourselves in that conspicuous place called the underground. These pictures are a pictorial corrido of that place. Who these people are and where they came from are marginalized by their common identity as Angelenos seeking oblivion from the reality that surrounds them.

# 27

# Apuntes para el fin de la ciudad más grande del mundo

Rogelio Villarreal Macías

El paisaje del nuevo milenio

Después de la caída del Muro de Berlín y del estrepitoso colapso de las economías planificadas de la Europa Oriental, no tardaron en aparecer los abanderados del libre mercado y de la democracia occidental para proclamar la victoria irrebatible de su ideal económico, arrogándose para sí la razón histórica ante el cadáver del Leviatán totalitario. En *El fin de la historia*,[1] Francis Fukuyama sostiene que, en adelante, el devenir de las sociedades libres y democráticas, al amparo de la mayor potencia del planeta, sería uno de constante desarrollo y bienestar general: la consumación del ideario de la modernidad. Ay de aquellos que se apartaran del airoso credo dictado por el Fondo Monetario Internacional y la Comunidad Económica Europea. Ay de quienes se atrevieran a cuestionar siquiera la nueva hegemonía mundial lidereada por Ronald Reagan y Margaret Thatcher: díganlo si no el fanfarrón dictador iraquí Saddam Hussein y el último dinosaurio comunista, Fidel Castro—suicidas prestos a desafiar al imperio en aras de una falsa dignidad y una independencia dudosa, en tanto sus pueblos sufren hambre, miseria, y opresión.

El panorama parecía ideal. Sin embargo, fue la misma historia, poco tiempo después, la que se encargó paradójicamente de contravenir las tesis del economista estadunidense. La violencia étnica y fundamentalista estalló con mayor fuerza que nunca en la constelación de naciones resultantes de la fragmentación de los Estados del socialismo real y en otros pueblos de Asia y África; los enfrentamientos interraciales y la violencia de la ultraderecha arreciaron en Los Ángeles, Nueva York, y otras urbes metropolitanas; la pobreza, la hambruna, y las enfermedades parecieron ensañarse aún más en los expoliados países tercermundistas. Por si fuera poco, la corrupción de los líderes

del neoliberalismo en todas las regiones del mundo rebasó con creces cualquier ficción cinematográfica.

En México, las promesas salinistas de modernidad y bienestar económico se convirtieron de la noche a la mañana en terribles mentiras. A la sombra del gobierno, funcionarios y empresarios amasaron en pocos años fortunas fabulosas en complicidad con los grandes capos del narcotráfico. En la calle, era difícil encontrar a alguien que no se sintiera afrentado, violentado. Grupos de indígenas chiapanecos se levantaban en armas el mismo día que entraba en vigor el Tratado de Libre Comercio entre México, Estados Unidos, y Canadá: nuestro fallido pase a la modernidad. El desengaño producía un crudo malestar que se acentuaba frente al cinismo y la miopía de las élites dirigentes, ocupadas en dirimir sus diferencias con peliculescos asesinatos a sangre fría: los casos de Colosio, Ruiz Massieu, y el Cardenal Posadas siguen envueltos aún en el más profundo misterio. El escenario de la degradada vida nacional es desde entonces un verdadero teledrama con abundantes dosis de sangre, intriga, corrupción, y esoterismo. De esa manera, inició nuestro abatido país el último tramo hacia el año 2000, hacia el futuro que los textos apocalípticos y algunos de ciencia ficción ubican precisamente en esta era.

**Nostalgia: la vuelta al pasado**

Como en el Renacimiento, los tiempos que corren se caracterizan, además, por sus frecuentes miradas críticas al pasado. La nostalgia es un componente esencial de la era actual; es la materia prima con la que se moldea el futuro instantáneo o, si se quiere, este presente eterno apuntalado por un torrencial y caprichoso flujo de referencias y reminiscencias.

La desbocada carrera del marxismo político y del liberalismo económico hacia el progreso, hacia el último y paradisiaco estadio de la humanidad, fue interrumpida bruscamente por la posmodernidad con la furia de un dios mitológico enardecido ante la estupidez crónica de los mortales: los escombros del comunismo aún dan la vuelta al planeta y la barbarie, la hambruna y la muerte oscurecen el paisaje idílico de la economía global: no sólo en la periferia de las metrópolis sino en sus mismas entrañas, la miseria y el desasosiego desmienten con hiriente contundencia la victoria del así llamado «mundo libre.» Lejos del bienestar prescrito con rigidez bíblica por los postulados marxistas y del libre mercado, los signos y los presagios actuales parecen indicar que el mundo se acerca cada vez más a la catástrofe. A pesar de los intentos de organismos civiles, de derechos humanos y pacifistas en casi todos los países, es posible que la humanidad no llegue a conocer nunca mejores condiciones de vida ni una civilización igualitaria y justa—ora sí que ya ni llorar es bueno. . . .

Pero la violencia ha ido siempre de la mano de la historia, y por lo mismo, en todas las épocas, los hombres han vuelto la mirada y su pensamiento al pasado tratando de esquivar las penurias y los desastres, las guerras y las invasiones de su presente. Ironías de la propia historia si vemos que nunca ha habido un lapso sin conflictos ni desastres de todas las magnitudes en algún punto del orbe. Sin embargo, nunca como ahora había sido tan pronunciada la vieja sentencia «todo tiempo pasado fue mejor»: un suspiro en labios de gente perpleja, desesperanzada en el mejor de los casos, y una oración desesperada para pueblos enteros devastados, inermes, al borde del exterminio.

La nostalgia y la idea renacentista de la refundación de las sociedades son un elemento esencial de la ecuación que rige nuestros destinos, como lo demuestra la memoria que alimenta casi cada acto de nuestra cotidianidad: la inocencia de la infancia, los amigos muertos, los primeros amores: momentos a los que se vuelve reiteradamente.[2]

### La extinción de las revoluciones

El tiempo de las grandes revoluciones ha pasado a la historia, señalan Octavio Paz y Ryszard Kapuszinsky, y apuntan que éstas han cedido el paso a pequeñas revueltas pragmáticas, a movimientos regionales de reivindicaciones inmediatas. Los pueblos quieren vivir bien y en paz; sólo quienes reemplazaron las grandes utopías con fanatismos ideológicos, económicos, o religiosos propalan y defienden a sangre y fuego su intolerante y mezquina visión del mundo. Pero unos y otros tienen su propia versión del pasado añorado, quizá un pasado remoto—el *illo tempore* de Eliade y Lévi-Strauss—en el que no había extraños ni disidentes ni, por ende, amenaza que temer: un tiempo mítico de armonía primigenia.

Los habitantes de la Ciudad de México anhelan los tiempos—de los que hablan nuestros padres y abuelos—en que podían transitar sin miedo, en que conciliaban el sueño sabiendo que nadie osaría irrumpir en sus casas para violar y sacrificar a las mujeres y a los niños y apropiarse de los bienes ajenos. Extrañan la época en que los gobernantes robaban sin tanto descaro y los crímenes de Estado no eran tan cínicos y algunos, en ocasiones, no quedaban impunes.

En los fatídicos tiempos que vivimos, la nostalgia se ha vuelto un recurso vital. Pensar en un pasado menos escabroso que el presente puede llegar a convertirse en un breve, íntimo acto de subversión. Acaso la poderosa evocación de tiempos en que aún se respetaban valores como la vida y la integridad personal contribuyeron a la victoria electoral de un político como Cuauhtémoc Cárdenas, que aparecía honesto y capaz a los ojos de una ciudadanía burlada por la dinastía priista, corrupta, y violenta, y que amenazaba eternizarse en el poder.

El recrudecimiento de la violencia cotidiana en la Ciudad de México y en el resto del país tiene en los habitantes de la calle a sus víctimas por excelencia. Lo que antes eran incidentes aislados ahora son asaltos, asesinatos, y violaciones a la vuelta de la esquina. Por si no bastaran las constantes humillaciones y afrentas del régimen, numerosas hordas de criminales—hijos del sistema que

prometía «bienestar para la familia»—se desparraman por todos los rumbos de la metrópoli asolando a la población. Abordar un taxi se convierte en una aventura que puede ser fatal y así lo advierten las embajadas extranjeras, provocando la indignación de la oficialidad y hasta de la oposición vehementemente antiintervencionistas. (Una situación parecida se vive en las grandes ciudades de Estados Unidos, el país que tiene casi en exclusiva la patente de los más temibles asesinos seriales. Psicópatas y francotiradores se reproducen en proporciones escalofriantes volviéndose antihéroes de la noche a la mañana mientras estudiantes y clientes de McDonald's caen como moscas. Las milicias ultraderechistas amenazan con la guerra al Estado y, de no haber sido aprehendido, el *unabomber* estaría tramando su próximo atentado. El lado oscuro del American Way of Life es una mancha que se extiende como un virus fuera de control.)

Si bien la violencia en cualquiera de sus formas—la coerción estatal, la intimidación militar, la del último de los ladrones— nos alcanza a todos, no es menos cierto que los agravios más crueles y virulentos se han cebado con especial encono en las mujeres. Las páginas de las secciones policiacas dan cuenta de crímenes que rebasan el colmo de la insania: una muchacha es secuestrada y robada en su propio auto y, para liberarla, los

delincuentes le exigen escoger entre violarla o mutilarle los pezones: la víctima aterrorizada escoge esta última salida. Otra estudiante es secuestrada a la salida del metro Auditorio y asesinada a puñaladas al intentar resistirse a la violación tumultuaria a manos de una pandilla que sólo quería divertirse pero, como declararon a la prensa, se les pasó la mano . . .

En *The Chalice and the Blade: Our History, Our Future,*[3] Riane Eisler sostiene una atractiva tesis sobre la violencia inmanente de las sociedades antiguas y modernas (y que en algunos pasajes recuerda un poco los tempranos escritos de Engels sobre las primeras formas de organización social). Las sociedades primitivas del neolítico, dice Eisler, vivían en un equilibrio más o menos estable debido a la participación igualitaria de hombres y mujeres en las distintas labores de la comunidad. La mujer era venerada entonces como dadora de vida, y la deidad más importante era la Gran Madre Tierra. Estas civilizaciones llegaron a alcanzar altos grados de sofisticación, como lo demuestran las investigaciones arqueológicas de Catal Huyuk, en la antigua Anatolia, y en la isla de Creta. Sin embargo, pueblos más primitivos que ensalzaban la fuerza masculina, así como la destreza en actividades como la caza y la guerra, habitantes de las zonas periféricas del planeta, comenzaron a invadir paulatinamente a las sociedades más avanzadas, provocando un cambio radical en su sencilla y armónica estructura social. La sociedad solidaria sufrió un dramático viraje hacia una basada en la acumulación, la esclavitud, el sometimiento de la mujer, y las guerras de conquista. La sociedad dominante, bajo la hegemonía masculina, se sofisticó hasta llegar a las variantes de distintos signos ideológicos en que hemos vivido y crecido.

No obstante, Eisler advierte en la proliferación mundial de organizaciones feministas y de derechos humanos claros indicios que llevarán, algún día, al restablecimiento del arcaico equilibrio perdido, a la refundación de la sociedad sobre nuevas bases de cooperación y armonía.

## Vivir y morir en la Ciudad de México

«La violencia es la partera de la historia,» sentenciaba a mediados del siglo pasado un viejo filósofo judío alemán al que nadie cita ya si no es para denostarlo. Pero, aunque se refería específicamente a la violencia revolucionaria, no puede negarse que la violencia en todas sus formas y gradaciones es el signo de nuestra resquebrajada/convulsa/sombría historia contemporánea.

El país está enfermo, y nada parece indicar ni el más leve asomo de mejoría, a pesar del optimismo hueco de las declaraciones oficiales. La descomposición del sistema y el recrudecimiento de la crisis ha generado la peor cascada de violencia e intolerancia de su historia reciente—magnificada y transmitida al instante por los medios de comunicación desde cualquier punto del planeta— a pesar de los demagógicos e inadmisibles discursos en contrario de funcionarios y gobernantes. Las actitudes xenofóbicas y la paranoia de la clase política se ensañan más a últimas fechas con los observadores extranjeros y mexicanos de la situación de los derechos humanos en Chiapas, así como con las minorías raciales, sexuales, y políticas.

Nunca como en nuestros días la Ciudad de México se había visto tan amenazada por la destrucción y la debacle, nunca antes se había experimentado tal estado de zozobra y angustia. El imponente volcán que vigila el Valle de México escupe humo y cenizas mientras a sus pies, 20,000 niños trabajan y duermen en la calle, víctimas de los abusos y la extorsión policiaca. Miles de indios y campesinos arriban diariamente a las estaciones de autobuses buscando un mendrugo para comer; miles de prostitutas y travestis recorren las calles en busca de un poco de dinero. El otrora límpido cielo del Anáhuac es un espeso manto gris que se cierne como una maldición sobre nuestras cabezas. Los bosques se incendian ante el estupor y la ineptitud de las autoridades. Las fábricas y los rascacielos se multiplican como hongos perniciosos, devorando kilómetros de campos y montes y asfixiando incluso las escasas zonas arqueológicas que aún perviven dentro del área metropolitana. La mancha urbana se extiende más allá de donde alcanza la vista y sus límites se borran nadie sabe dónde. La sangre y el terror son los guardianes más efectivos de una ciudad en estado de sitio. La tragicomedia que vive el país, al igual que las telenovelas de la televisión mexicana, aún da para muchos capítulos más.

Nadie está libre de violencia. De ejercerla o de sufrirla. Aun los estados más democráticos recurren a ella para restablecer el orden y la calma. Los poderosos golpean y vejan a los débiles, los patrones a los obreros, los hombres a las mujeres, los padres a sus hijos, los policías a los ciudadanos, las guardias blancas a los campesinos, los machos a los homosexuales, los ladrones a los transeúntes, los católicos a los evangelistas—y viceversa. La violencia llena de cicatrices vergonzantes la piel de los países del mundo. La violencia se aloja en nuestros propios hogares, y en la calle todos somos enemigos potenciales a los ojos del otro. La violencia sólo podrá conjurarse si aprendemos a desarmar el mecanismo que la activa.

Pero, como la historia ha dejado de ser lineal y progresiva, quién dice que algún día no puedan volver los viejos buenos tiempos. . .

## Notas

1. Francis Fukuyama, *The End of History* (Nueva York: Free Press, 1992).

2. Dos filmes ya emblemáticos tienen a Los Ángeles, la ciudad del futuro, como entorno a un tiempo vital y avasallante. Rachel, la hermosa replicante de *Blade Runner* (Ridley Scott, 1982), se humaniza precisamente gracias a los recuerdos de otra persona implantados en su memoria. En *Días extraños* (Kathryn Bigelow, 1995), cinta que anticipa la apoteósica entrada del nuevo milenio, un astuto ex policía trafica clandestinamente con climáticas experiencias personales grabadas mediante un dispositivo que convierte los ojos del testigo-actor y cómplice en nerviosas cámaras de video y el cerebro en una puntual videograbadora. Lenny Nero, el traficante, repasa obsesivamente la cinta que reproduce los momentos de felicidad vividos con su ex novia. El dispositivo reproductor permite, más que evocar, experimentar de nuevo con fiel intensidad los acontecimientos vividos y registrados, así sean los de otro individuo al que nunca se conocerá.

3. Riane Eisler, *The Chalice and the Blade: Our History, Our Future* (Nueva York: Harper and Row, 1987).

# CHAPTER 28

# The Case of Walter Betancourt
# Reflections on Identity and a Sometime Angeleño

John A. Loomis

*Our living depends on our ability to conceptualize alternatives, often impoverished. Theorizing about this experience aesthetically, critically is an agenda for radical cultural practice. For me this space of radical openness is a margin—a profound edge. Locating oneself there is difficult yet necessary. . . . Marginality nourishes one's capacity to resist. It offers the possibility of radical perspectives from which to see and create, to imagine alternatives, new worlds.*

—bell hooks[1]

This chapter is not about Latino Los Angeles. It is not about Los Angeles. The protagonist here, Walter Betancourt, did live briefly in Los Angeles from 1957 to 1960, but that is marginally relevant. But then we are discussing the importance of margins, are we not? Betancourt left no discernible mark in Los Angeles, and it is unclear what mark the city left on him. But this chapter is about the nature of the margin as the creative locus of identification and proposes to raise issues concerning identity that might provide other perspectives from which to regard Los Angeles and its evolving identity, because Walter Betancourt's creative life and work were inextricably caught up in issues of identity long before identity became an issue.

Walter Betancourt was born July 18, 1932, in New York City. His grandparents, people of modest means, had emigrated from Cuba to Tampa at the time of the Cuban War for Independence.

AUTHOR'S NOTE: Research has been assisted through the generosity of the Getty Research Institute, the Associación Hermanos Saíz of the Cuban Ministry of Culture, and the Unión Juventud Comunista de Cuba. I am also indebted to the assistance of Gilberto Seguí Diviñó, Rosendo Mesías, Eduardo Luis Rodríguez, Roberto Serge, the late Arturo Duque de Estrada y Riera, and the late Julia María Leonor Fernández Bulnes de Betancourt.

Through their hard work, and that of his parents, the family had risen up the economic ladder and achieved the American Dream by the time of his birth. Betancourt grew up in the cosmopolitan comfort of a solid urban middle-class family. Family vacations to Cuba served to connect him to his heritage, but by and large, he lived an "American" existence. He studied architecture at the University of Virginia, bastion of Anglo American identity, graduating in 1956. In that same year, he served a brief tour of duty in the U.S. Navy, stationed at Guantánamo, where he witnessed from afar the Moncada Uprising of the July 26th Movement, the beginnings of the Cuban Revolution. The following year, he moved to Los Angeles to work for Richard Neutra, whom he admired both as a designer and as a person of progressive social commitment. But the reality of Neutra's office, where he worked without pay, did not meet his expectations or ideals, and he left after six months. Betancourt stayed on in Los Angeles with his young wife, Leonor, while in Cuba the Revolution gained momentum. It also gained support from abroad, and Betancourt participated in solidarity committees as he continued to develop his professional skills in the offices of John Lautner and others and take postgraduate courses at UCLA. It is clear that the events in Cuba and his growing disenchantment with Los Angeles and the United States were having a profound effect on his own sense of identity. In 1959, he interviewed with Frank Lloyd Wright at Taliesin and turned down what he otherwise would have considered an ideal opportunity, an offer to work under Wright himself. Instead, he made a critical decision—to go to Cuba and dedicate his design skills to the newborn Cuban Revolution.

What did this act mean, and what are we to make of Betancourt's few years in Los Angeles? The city was a place where people came to create and re-create their identities. This was standard fare in the film industry, but it was true in other fields as well. Both Rudolf Schindler and Richard Neutra left the formally restrictive society of old world Vienna to reidentify themselves and their talents in the artistic freedom that was Los Angeles. Frank Lloyd Wright turned a new stylistic chapter in his work with his exploration of pre-Columbian identities in the work of his Los Angeles period. Later came an influx of intellectual émigrés fleeing Nazism, who would leave their mark on the culture of Los Angeles. But that was a different era. When Walter Betancourt arrived, Los Angeles was no longer the bohemian avant-gardist environment it had been in the prewar era. It had itself gone from being a margin to a center. Several things happened to Betancourt during the late 1950s in Los Angeles. He grew and gained experience as a professional and became more confident of his design abilities. His professional idealism convinced him of the value of the art of architecture and of the responsibility of the architect as primary guarantor of a work's cultural value. But he became profoundly disillusioned with architecture as practiced under capitalism. He experienced a political awakening that coincided with soul searching concerning his own identity and ethnicity. This identity, as an "American," became challenged and began to deconstruct and reconstruct itself. The Cuban Revolution was the catalyst that brought this about. Walter Betancourt's move to Cuba was a primary act of identification.

Walter Betancourt arrived in Havana on August 8, 1961, when the young Revolution was still in a state of euphoric bacchanal. Betancourt sensed, however, the forthcoming doctrinaire tendencies of the Revolution that would soon restrict architectural development, however, so he decided to move far from Havana's ideological center.[2] He practiced a politics of place that took him to Cuba's eastern provinces, Holguín, and finally Santiago, where he discovered the importance of being *oriente*. Santiago and eastern Cuba have historically been resistant to Havana's center and have always asserted a certain degree of political and cultural independence, something that the revolutionary government respected. In terms of marginality, Betancourt sought out margins within margins where he conducted a critical counterhegemonic discourse through architecture. Cuba was marginal to the United States and the rest of the developed world. Santiago was marginal to Havana. And the deeply rural locus of his two most important works were marginal to Santiago itself. The counterhegemonic practice Betancourt established was counter to architecture as practiced under capitalism in general and counter to the hegemony of its practice in the United States in particular. Ironically, however, it was also counter to the hegemony of the prevailing norms of architectural practice in Cuba. And for this reason, despite its significance, Betancourt's work has by and large lacked recognition by the architectural establishment of his adopted country. By 1963, private architectural practice was abolished in Cuba, and the Colegio de Arquitectos was closed. Architects were now expected to serve primarily as technicians, part of a team of engineers who would resolve Cuba's many building needs through massive industrialized solutions based on Soviet models. That Walter Betancourt was able to survive and thrive under this situation is truly remarkable, a testament to both the power of marginality to nourish the capacity for creative resistance and Betancourt's own charmingly persuasive personality. Moreover, idealist that he was, Betancourt chose to live an ascetic existence, a true communist, marked by sacrifice and self-denial, that placed him personally beyond criticism from party hardliners.

Forestry Research Lab, Guisa

In Betancourt's relatively short but productive life in Cuba (he died at forty-six in 1978), he is credited with fifteen built works and more than thirty unbuilt projects. These works stand as examples of an architecture of critical resistance and a multilayered approach toward constructing identity that relies on a process of cultural hybridity and syncretism. Wedded to this process was a strong reference to and reliance on the work of Frank Lloyd Wright. Wright's appeal to Betancourt is not only evident stylistically but also understandable ideologically when one considers Wright's own position as a perpetual anticentrist. In his chapter on critical regionalism, Kenneth Frampton notes an interesting parallel regionalist development of a similar Wrightian tendency in the Italian Alps of Ticino in the mid-1950s when there was a conscious attempt to establish an organic regional alternative to rationalist modernism.[3] Betancourt's primary homage to Wright is the Forestry Research Laboratory at Guisa (1970). High on a mountainous site, this Taliesin-like complex arranges itself

Forestry Research Lab, Guisa

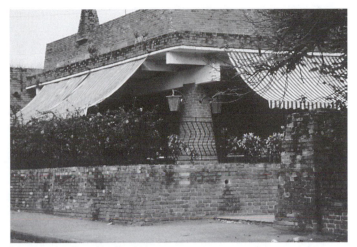

Restaurant Las Pyramides, Santiago

across the topography, accommodating the contours to avoid any cut and fill, and embracing the vegetation so that no trees were removed. The project affirms his strong conviction in building the site as the prerequisite for constructing identity, and no site could have been more marginal, located as it is in the remote reaches of the Sierra Maestra, birthplace of the Cuban Revolution.

Part of Betancourt's identification through hybridization drew also from the vernacular. He had a great respect for both vernacular form and constructive tradition. Both these positions ran counter to the tenets of the centralized Ministry of Construction, which regarded vernacular precedents as "romantic, folkloric and derivative of residual bourgeois ideology" and looked on traditional building techniques as "backward holdovers from underdevelopment." The assimilated and interpreted form of the *bohío*, the typical hut of the Cuban peasant, is evident in the pavilions of the otherwise Wrightian Forestry Laboratory at Guisa. Yet Betancourt's position regarding this architecture of everyday life was not populist because he believed not merely in the appropriation but also in the transformation and reinterpretation of vernacular form through the hands of the architect. Also worth noting are the expressive wood framing and rafters that delineate the walkways and the heavily articulated brick walls of the complex. This love for the expressive nature of brick, which has deep Spanish origins, is evident also in the restaurant Las Pyramides (1966), in a modest residential neighborhood in Santiago. Here the brickwork presents multiple readings that recall both pre-Columbian monumental form and Mies van der Rohe's monument to communist martyrs Rosa Luxemburg and Karl Liebknecht. The "tactile and tectonic" qualities of Betancourt's work often dominate over the "visual and graphic."[4]

This is nowhere more valid than in the Cultural Center of Velasco (1964-1991), where brick, concrete panel, terra-cotta tile, plaster, iron, and wood combine in the highly sensory composition and expression of cultural identity. Up until the establishment of this arts center, this small provincial town was known

Cultural Center, Velasco

for little more than the beans produced in its nearby fields. In no other project of Betancourt's is the expressive potential of brick construction as a generator of form more evident than here. Part of the credit goes to the remarkable partnership Betancourt struck with a Spanish master mason, Nicasio Santana, who had fled Spain for Cuba in the early years of the civil war, having refused to serve in Franco's army in Morocco. Together, the architect and the master builder created a complex that departed radically from Betancourt's Wrightian inclinations. At Velasco, the derivation of form from the constructive process results in a project of great episodic poetry evocative of Spanish tradition and Caribbean spirit. It was a project that took twenty-seven years and was finished long after the deaths of both Betancourt and Santana, thanks to the perseverance of Betancourt's associate architect Gilberto Seguí Diviñó. The project was also identified by the enthusiastic support of this community of farmers and artisans. Yet it would be a mistake to say that the design process was participatory in the sense that it is commonly defined. Betancourt, with Santana, maintained a firm hand on the formal development of the complex, while nevertheless responding to the community's needs

Cultural Center, Velasco

and desires. The Cultural Center of Velasco provides a formal and symbolic identity that has become a source of local pride to this otherwise typical poor rural Cuban village. In a country in which so many of the public works suffer from neglect and a lack of maintenance, the Cultural Center of Velasco is always kept tidy and in good repair by the volunteer efforts of a community that identifies with its well-being.

But despite local appreciation of his work, Walter Betancourt's architecture has been virtually unknown in Cuba up until 1992, when through the efforts of Gilberto Seguí Diviñó, Eduardo Luis Rodríguez, Rosendo Mesías González, and others, a small exhibit of his work was organized to coincide with the Bienal in Havana. Although the exhibit received support from the Ministry of Culture, the Ministry of Construction and the Union of Architects and Engineers were decidedly absent in their support. For these central authorities, charged with overseeing the country's construction needs, Cuban revolutionary identity was embodied by functionally and technically determined projects that were repeated on a massive scale with little or no consideration to site and local conditions. The quality and marginality of Betancourt's architecture presented an uncomfortable challenge to the "one correct line" official mentality.

Walter Betancourt's example is a rather compelling, perhaps even disturbing, challenge to those of us who write articles, participate in conferences, design projects, scramble to get them published, and are dependent on a system of rewards and recognition. His example suggests that one might just take leave of it all, retreat to an isolated place, turn inward, and quietly make one's mark, disregarding the rest of the world, not caring for recognition, confident in only the nature of the concrete creative act of—identification.

*This is an intervention. A message from that space in the margin that is a site of creativity and power, that inclusive space where we recover ourselves. . . . Marginality is the space of resistance. Enter that space. Let us meet there.*

—bell hooks[5]

Cultural Center, Velasco

## Notes

1. bell hooks, *Yearnings: Race, Gender, and Cultural Politics* (London: Turnaround, 1991), 149.

2. For a further discussion of architecture under the Cuban Revolution, see John A. Loomis, "Architecture or Revolution? The Cuban Experiment," *Design Book Review* (summer 1994): 71-80. For further information about Walter Betancourt, see John A. Loomis, "The Architecture of Walter Betancourt's Quiet Revolution," *Progressive Architecture* (April 1995): 41-44; Roberto Segre, *America Latina fim de milénio,* Sao Paolo: Studio Nobel (1991); and Gilberto Seguí Diviño, "Stazione Forestole Sperimentale," *Architectura Croniche e Storia* (April, 1995): 272-5.

3. Kenneth Frampton, *Modern Architecture: A Critical History* (New York: Thames and Hudson, 1992): 322.

4. Ibid., 317-18, for a discussion on the presence of the "tactile and tectonic" in the architecture of critical regionalism.

5. hooks, *Yearnings,* 152.

# Learning from East L.A.

Robert Alexander González

There is a perversity in the learning process: We look backward at history and tradition to go forward; we can also look downward to go upward. And withholding judgment may be used as a tool to make later judgment more sensitive. This is a way of learning from everything.

from *Learning from Las Vegas*[1]

In *Learning from Las Vegas,* Venturi, Scott Brown, and Izenour singled out the anomalous and dynamic urban and architectural transformations of Las Vegas. They looked at commercial vernacular and urban sprawl to systematically analyze the city and formulate theories about how architectural configurations can communicate certain messages. "Learning from East L.A." should not suggest a repetition of this "downward to go upward" method, which can be criticized for theorizing "low-culture" for high-cultural production. Instead, we might consider what a study of East Los Angeles can offer emerging Latino cultural landscapes in longitudinal and latitudinal directions, as well as in parallel and unifying ways. Although it might seem contrived to strategize the formation of whole new Latino cultural landscapes, active explorations of such possibilities already exist. A paradigm of a

Latino built form has recently been proposed by a number of architects to counter the colonial-related aesthetics that have traditionally served as "stand-ins" for this ethnic group, specifically aesthetics that came from the Mission- and Spanish-style palettes.

What can East L.A. really teach us about alternative and representational paradigms? Let us look closely at three interpreta-

tions of the word *latitude,* which can mean *freedom from limitations; extent, range;* and *the angular distance north or south of the equator.* These terms help us diagnose the forces that have allowed East L.A.'s experimental urbanity to take form. Systematic analyses of these forces may help future Latino cultural landscapes shake loose the design impasse that is colonization's legacy. Let us revisit how these inhibiting conditions emerged and where they stand today. Consider a theory of *Latitunidad,* a strategy for interrogating immutable Latino landscapes.

### 1. freedom from limitations.

Specifying Ramona

Consider first *freedom from limitations.* It is seen in East L.A.'s unconventional aesthetic of contemporary urbanity. Here, gang tag stylizations are not whitewashed away by municipal authorities; instead, palimpsest walls are reworked again and again until complex new strains of urban calligraphy are invented. Similarly, low-rider imagery emerges from East L.A. garages and appears in weekend ritualized car promenades, on Internet cybersites, and in low-rider magazines. Gripping murals tell the tale without ignoring the trials and tribulations of urban life, and community members don't always appear in the typecast roles found in older murals. This segment of Latino life in L.A. is about the subversion of the stereotypical. These qualities represent an important stage in the decolonization of the Latino aesthetic.

Compare this unique L.A. patina with the colonial trappings that continue to represent *latinidad* in most U.S. cities, and the images that limit this ethnic group elsewhere begin to surface. Anglo American appropriations of indigenous and Spanish colonial aesthetics, the favored pair, have consumed and limited architectural discourse by permanently defining this cultural landscape through a Hispanic colonial past. They have complicated the built environment by establishing historical postures as es-

sential elements of the Latino cultural landscape. Indigenous and Spanish revival architecture, in all its variations, was mostly the result of cultural production by and for Anglo Americans, but, today, these trappings commingle with the Latino community's own expressions of resistance, producing confusing results. These colonial aesthetics are certainly a part of East L.A., but of import here is that they yield ground to alternative paradigms.

The chapters that compose this volume demonstrate East L.A.'s revolutionary transformation of colonial aesthetic traditions. The problematic that the two aesthetics pose, however, warrants further exploration. How did these historical citations develop, and what continues to sustain them both "outside" and "inside" Latino communities beyond L.A.?

First, Renato Rosaldo's notion of "imperialist nostalgia," the process whereby "agents of colonialism long for the very forms of life they intentionally altered or destroyed," helps explain why the indigenous aesthetic was developed by Anglo Americans.[2] The literature that locates these imperialist practices is too vast to cover here, but some key figures and historical developments should be mentioned. Ancient indigenous architectural commodification dates to the mid-nineteenth century, to the travel accounts and illustrations of explorer John Lloyd Stephens and architect Frederick Catherwood. Later, the physical objectification of Mesoamerican architecture in the United States occurred most profoundly when architectural scenographies of "savages" were constructed in the late nineteenth and early twentieth centuries for the world fairs and expositions. Popularization happened again during the Mayan and Aztec revival period in the 1920s and 1930s, especially with the work of the Mayan revival booster, architect Robert B. Stacy-Judd. With regard to Los Angeles, one of the most famous cases of appropriation is found in the Mayan-influenced architecture of Frank Lloyd Wright, as seen in the Barnsdall or Hollyhock House (1918-1921). The spread of Wright's appropriation continues today. In a Donna Karan advertisement[3] for a new "occult"-like fashion line, a sultry model ina dark cape is photographed against a Mayan-inspired Wright building (Ennis House, 1924). The model's disposition and the indigenous imageryinthe background push all the stereotype buttons.

Stereotyping is further complicated when these moments of "imperialist nostalgia" are juxtaposed with the use of the same imagery by Latino communities. In Laredo, Texas, for example, the city's poorest *barrio* was renamed El Azteca in the 1940s after the nearby downtown Aztec Theater. This gesture was not an indication of a raised consciousness among the local Latino population; it was simply a matter of proximity. The neighborhood's citizens eventually thought differently, however, when they assumed the indigenous reference as a representation of their struggles as Mexican Americans. What do we make of these dual forms of "outside" and "inside" appropriation? Does the use of such imagery by some urban or Laredo's Latinos confirm the thinking behind the imperialist gaze? How do we confront the coexistence of the same imagery for both appropriating and resisting practices, especially in light of creative production?

The second type of Anglo American appropriation is the Spanish aesthetic of Mission revivalism and restoration. This also presents a double bind, especially in consideration of Anglo and Latino participation in Mission restoration and boosterism. Restoration was prompted, in part, by the vast popularity of Helen Hunt Jackson's best-selling novel *Ramona* (1884). This novel was set in the California of the Californios, when they were being displaced between 1848 and the close of the century by a new Anglo ruling class. It was through Ramona's character that Anglo American appropriation was visualized—first by a small audience interested in restorations—later in the century, as the novel

continued to sell, by a growing number of west coast suburbanites who desired the Mission revival home. The proliferation of the Mission aesthetic through restoration and revivalism, however, never presented an accurate portrayal of mission life. I am referring here to the power and control that the Spanish *padres* exerted on Native Americans under the Spanish mission system between 1769 and 1834.

The twenty-one California missions resonate strongly with Foucault's analysis of eighteenth-century means of correct training and his analysis of Bentham's Panoptican, when we consider the typical mission quadrangle where the "spatial 'nesting' of hierarchical surveillance" took place.[4] But this was completely left out of the imagery customized for a new Anglo American public. David Hurst Thomas has criticized the fact that "thousands of tourists flock annually to California's missions, luxuriant amidst the jasmine and ever-blooming lantana," because, as he explains, "the unfortunate truth is that these cornucopian mission gardens are pure *Ramona*-derived hyperbole. Period paintings, textual descriptions, photographs, and archaeology amply demonstrate that such flowery enchantment never existed in the original mission."[5] The barren dirt quadrangles Thomas refers to were, actually, more like internment camps than lush gardens. This false imagery, which is perpetuated by numerous festivities across the Southwest, confounds the Latino cultural landscape. The problem of appropriation is further complicated when ethnicity is rendered invisible in the process. Strong cultural ownership is implied as a result, not so much in a case like the novel *Ramona* but in the popularization that can evolve from it, as in the *Ramona*-like fiestas and others that celebrate a colonial past. These fiestas do not concentrate on how ethnic groups create and celebrate their own identity. Instead, they focus on how the dominant culture can reinvent ethnicity. The act of claiming "the Wright style" resembles other forms of cultural ownership by the dominant culture, such as the practice of specifying "Ramona's adobe," which was common during the proliferation of Mission revivalism, or the marketing of "Cliff May's Ranch Style Houses" by *Sunset* magazine in the late 1930s and 1940s. The ethnicity originally associated with the architecture is dropped entirely from the discourse.

The Latino community's participation in Spanish colonial boosterism causes further complications. Lisbeth Haas has discussed how the Club Hispano Californio and the Mission Indian Federation were involved in mission restoration after secularization. She explains that Californios became interested in "who would control the labor of Indians after they were freed from their obligations to the mission." The Californios' "interpretation of freedom and definition of their territorial identity, based as these were on colonial law, remained closely tied to the ideas that had sustained the colonial regime."[6] This is a tenuous issue because the development of the Mission aesthetic paralleled the alignment by *mestizos* or Mexican Americans who passed for Spaniards and Californios (because of their lighter complexion) with the mission system's underlying power structures. This alignment allowed Mexican Americans to claim some type of superiority over Native Americans. Mission-style overindulgence has since resulted throughout the Southwest, and its underlying meaning has rarely been criticized by Latinos. Instead, the Alamo's profile has become a ubiquitous architectural element exhaustively repeated throughout the state of Texas, and mission-related festivities in Texas, New Mexico, and California have become unifying celebrations for many Latino communities.

## 2. extent, range.

La Casa, La Causa, La Casita

*Extent and range* is approximated in East L.A.'s ad hoc barrio constructions and front yard shrines and in the unconventional urban props, the ubiquitous *taco trocas*. The entire L.A. area has been recognized for its extremist quality. Charles Jencks uses the word *heterotopolis* (love of difference) to describe L.A., but this has more to do with the creations by the Santa Monica School's "high style" architects (i.e., Moss, Rotondi, Mayne, and Israel) than with ethnicity. Jencks's study does not concentrate on East L.A.'s transformative and vernacular

*latinidad* or on Latino avant-garde architectural production. This latter point raises an important issue if we are to consider the entire extent and range of the Latino cultural landscape.

In the applied arts, literature, and music—the more accessible forms of cultural production—the Latino avant-garde has flourished successfully. Latino architecture, landscape architecture, and urban design have lagged behind these forms of expression. Latino/a architects have yet to make a name for themselves in the United States. Instead, imported Latin American architecture, such as the work of Mexican Ricardo Legoretta, is more likely to stand in for the Latinos in cities such as San Antonio, Los Angeles, and San Francisco. The Spanish architect Rafael Moneo was recently awarded the commission to rebuild L.A.'s Saint Vibiana Cathedral. Few cases of architectural and urban design trajectories that focus on *latinidad* exist. The urban designers called Barrio Planners are known for their efforts since the civil rights movement to influence urban design with a distinct Latino perspective, but an example such as this one is rare.

How did this present impasse in the environmental design fields occur? The impasse dates to the 1848 Treaty of Guadalupe Hidalgo, the official sale of vast northern Mexican lands to the United States. Mexicans who chose to stay in these transferred territories quickly encountered a rapid downturn of their economic fortunes. In Los Angeles County, for example, only 5 percent of Mexican American landowners managed to hold onto their property ten years after California gained statehood, despite the guarantees of the treaty. One consequence of this rapid loss was uncertainty and visible change. Architectural production from 1848 onward "reflected the shift in taste toward American design following the Yankee takeover. Adobe increasingly gave way to wood and fired-clay brick as new buildings began to resemble those of the eastern U.S."[7] The ruins of the *haciendas, rancherías,* and missions throughout the Southwest became a fixed representation of a Hispanic past for Anglo American newcomers. Meanwhile, real history was rewritten. Poverty-ridden barrios were romanticized and frozen in time by popular Anglo journals and travel accounts. In a less positive vein, these communities were also depicted as evidence of a stagnant Latin America. For example, Charles Fletcher Lummis renamed New Mexico the "National Rip Van Winkle" in his book *The Land of Poco Tiempo* (1891). "Poco Tiempo" was "Mexican Time," a temporal representation of a perpetually dormant cultural landscape.

A growing detachment from land also intensified this design impasse. The fate of the Mexican Americans was controlled early on by the demands imposed by growing Anglo American agribusinesses and railroad companies that, among other detriments, created a highly migratory people of the Mexican Americans. Deportation and importation, at the whim of the U.S. government, only aggravated this condition. For Mexican Americans, this period of detachment from land also marked a significant shift from the production of *home building* to the exploration of *homeland*. The social positions of Mexican Americans dropped as their earning capacity decreased. Consequently, they built less for themselves and instead to fulfill the aspirations of Euro-American city visionaries. This occurred at precisely the time that their relatives across the newly legalized border were engaged in nationalist movements that explored new building paradigms. Mexican Americans now faced a new self-identity that was distinct from the nationalist discourses in Mexico. The result was an experience that in many cases encompassed dual nationalist sentiments, and this has since shaped the collective identity of many Mexican Americans. The emergence of more inner-city barrios

brought on by the Great Depression and two world wars also detached Mexican Americans from their land: by 1930, more than 50 percent of U.S. Latinos lived in urban communities; by 1960, 80 percent of Mexican Americans lived in cities.

Mexican American communities soon concentrated their creative forces on emerging political agendas. The growing political climate of the civil rights movement inspired many Mexican Americans to identify with Cesar Chavez's struggle to unionize California farmworkers. Artist Esteban Villa, from the Mexican American Liberation Art Front, remembers the moment when the home became a politicized expression: a *casa* on Twenty-fourth Street in Oakland, California, was symbolically renamed *La Causa*, "the cause," in response to this political climate.[8] The symbolic transformation of *casa* to *causa* is important because it inaugurated a collective effort to actively produce artistic work, which is in direct opposition to the dominant culture. Tomás Ybarra-Frausto explains that

> the primary aim of [the Liberation Art Front] project was to surmount strategies of containment by struggling to achieve self-determination on both the social and aesthetic planes. An urgent first task was to repudiate external visions and destroy entrenched literary and visual representations that focused on Mexican Americans as receptors rather than active generators of culture.[9]

In response to Kay Turner and Pat Jasper's article "La Causa, La Calle y La Esquina," Ybarra-Frausto also commented that "inside the home, in the yard, and on the street corner—throughout the barrio environment—a visual culture of accumulation and bold display [was] enunciated."[10] These observations demand clarification because they suggest that the aesthetic changes were all-encompassing. These modes of intervention, however, only tinkered with a physical infrastructure that remained largely Anglo American. Thus, although Chicano art and literature flourished, architectural explorations remained derivatives of conventional building practices, although they were quite subversive.

More important, the explorations operated (and still do) within the trope of appropriation, which can range from the highly ephemeral ("here today and gone tomorrow,") to the quasi-permanent, where full-blown ad hoc constructions render original building templates *almost* invisible. Although *La Causa* was successful in promoting work in the literary and visual fields, other shapers of the cultural landscape (i.e., architects, landscape architects, and urban designers) played a limited role in cultural production at this time, mainly because of the low number of Latinos/as in these professional fields.

In recent years, however, the extent and range of the Latino cultural landscape have shown a slight shift in latitude. These changes have occurred throughout the country as borrowed, transported, and retained notions of space and place are reworked by Latinos who share a proximity to a nearby homeland. Although some Anglo American-initiated interventions have included efforts to preserve remnants of this Latino homeland—such as Olvera Street in East L.A. and *La Villita* in San Antonio, Texas—Latino architectural interpretations that affirm their own ethnicity deserve more attention.

First, consider the well-known example of the Puerto Rican *casitas.* The "Nuyoricans" who build *casitas,* the brightly colored balloon-frame structures that mimic those found in Puerto Rico, present an architecture of resistance. Each such *casita* "is a source of pride and memory—it articulates and validates the . . . Puerto Rican identity in space" by "transforming fragmented and discontinuous urban landscapes into 'cultural forms with continuity.' " Yet despite their ability to unify Puerto Ricans, *casitas* are allusional. They refer to "the power of other places [that] everybody recognizes, feels good towards and can identify with." In other words, they refer to Puerto Rico.[11] The *casita*'s strength is tied to this building type's changing symbolic meaning abroad. If we compare these *casitas* to the three Puerto Rican homes exhibited in the 1998 Dream Houses Exhibition in the Bronx, New York, however, we find another level of creativity and a complete reinterpretation of tropical or Caribbean architecture.[12] Although Frances Aparicio and Susana Chávez-Silverman consider the *casitas* as exemplary forms of oppositional practices in the way they retropicalize New York's public realm, the "dream house" *casitas* offer oppositional design strategies and alternative paradigms. The architectural language they introduce offers new interpretations of symbology, color, metaphors, and myth.[13] Another example of a tropical architecture that also expands the language of the Latino cultural landscape in the same way as the dream house example is seen in the *casita* designed by the

Ecuadoran architect Carlos Zapata. Zapata re-creates an alternative tropicalized environment by avoiding all historical citations and by exploring new materials and configurations that are highly abstract and ephemeral.

Two mid-1990s design proposals for public buildings in San Antonio, Texas, further expanded the possibility of a Latino architecture, although both designs raised many eyebrows in a city that has been rather conservative about ethnic architecture. First, a design proposal for a public building that was to represent *mestizo regionalism* was presented by the architecture firm Kell, Muñoz, Widgowsky, along with a manifesto that explained the ideology behind this newly minted concept. This design, however, was soon lambasted by local newspapers and the city council, and the project was never realized. Second, the commission for the University of Texas at San Antonio campus extension in west San Antonio, where the city's largest Latino population resides, was awarded to the architect José Saldaña. The architect proposed a *rasquache* architectural building, which refers to the Chicano vernacular ad hocism that Ybarra-Frausto describes as an underdog aesthetic that is "rooted in resourcefulness and adaptability" and which "makes for syncretism, juxtaposition and integration."[14] Saldaña's building, which was eventually built, raises questions about the applicability of an ethnic vernacular language to a large-scale institutional building. Furthermore, his architecture seems to be more about an assemblage of clashing colorful materials and canonic forms than about the moniker with which he chose to describe it. The geometric nature of this building also relates more, with its proportions and scale, to the highway it faces above ground than to the community it serves at the street level. A better example of the juxtaposition and integration that Ybarra-Frausto refers to is seen in the recent Zahedi House designed by Venezuelan-born architect Monica Ponce de León and her partner, Nader Tehrani (Office dA). This house not only explores innovative materialities but also approximates a more accurate depiction of a *rasquache* architecture with its undulating, folded, and punched metal corrugated sheets that hang almost freely in front of this Massachusetts residence. The architects attempt to integrate a conventional material with a standard house in an unconventional way; this creates an entirely new architectural language.

### 3. north and south of the equator.

House on Mango Street, House of the Spirit

Consider the last element of the triad, the migration in the *north and south* directions and the reshaping and reclaiming of East L.A. that results. Although a state of forced migration can result in detachment from land, as it has been shown, it can also result

in the spontaneous creation and appropriation of whole landscapes. This is evident in the manner that East L.A. has traversed
outside East L.A. proper. An example of this is the transformation in the 1970s and 1980s of MacArthur Park, the present-day
Little Central America in L.A. The once clean-cut, bourgeoisie
park is now an urban center and a symbolic point of arrival for
many Latin American immigrants. It is surrounded by Latin
American consulates, ESL learning centers, and the hustle and
bustle of swap meets and other informal markets. Not surpris-
ingly, this transformation has prompted white flight to occur—
ironically even by Otis School of Art, the institution that once
spearheaded a beautification project in the 1970s that included
the building of ethnic monuments throughout the park.
MacArthur Park's dynamic transformation and programmatic
shift are clearly a direct result of the latitudinal directionality of
the hemisphere.

The constant north-south diaspora, and the never-ending traveling back and forth of cultural signifiers, has also given music,
literary, and visual cultural production from both Latin America
and the United States a unifying sensibility. Development in the
environmental design fields, however, does not parallel the fluid-
ity that other forms of expression and representation have cap-
tured. It is true that common denominators that are found
throughout Latin America, such as the *plaza,* the *patio,* and the
*corredor,* are the threads that link multiple Latino and Latin American cultural experiences. Still, performance artist Guillermo
Gómez-Peña's call for an awareness of America as a continent

and not a country awaits interpretation in the urban and architectural realms. I am not rallying for a Pan-American aesthetic, but simply that the constant trans-Latino experience can inform the cultural landscape as well. Gómez-Peña explains that "in the new typology, an emergent axis of influence might lead from Los Angeles to Mexico City and from there to Bogota, Lima, Buenos Aires, Managua, Barcelona, and back to the *barrio*. Artists [can] go back and forth between different landscapes of symbols, values, structures and styles, and/or operate within a 'third landscape' that encompasses both."[15] This last interpretation of *Latitunidad* suggests that the places that are produced by the environmental design fields can traverse landscapes and borders as easily as the cultural signifiers found in novels such as *The House on Mango Street* by Sandra Cisneros and *The House of the Spirits* by Isabelle Allende. These texts coexist in the collective memory of many Latinos/as, although they represent different forms of *latinidad*.

■

My interest here has not been to point to East L.A.'s urbanity in order to glean principles for widespread application. Instead, I have attempted to analyze the conditions of the broader Latino cultural landscape that keep creative developments from flourishing. I attempt to offer glimpses of some of the developments that show promise. We need to look more closely at urban centers such as San Antonio, Texas, where a reliance on the Alamo's iconography and a deference to a visible Mexican cityscape coexist and continue to shape the cultural landscape. Similarly, we need to be more critical of the strong heritage-boosterism found

throughout the Southwest, especially in Santa Fe's grand Fiesta, where the reconquest of Native American lands by the Spaniards continues to be the focus of the celebration. These rituals have strengthened some communities in identity formation and community unification, but their subtle inhibiting nature must be tackled. Colonial aesthetics offer little or no vision for the aesthetic and practical needs of a population that will soon constitute the largest ethnic majority.

## Notes

1. Robert Venturi, Denise Scott Brown, and Steven Izenour, *Learning from Las Vegas* (Cambridge: MIT Press, 1977), 3.

2. Renato Rosaldo, "Imperialist Nostalgia," *Representations* 26 (spring 1989): 107-8.

3. "The Wright Style," advertisement for Donna Karan, *Los Angeles* Magazine (October 1997): 60-5.

4. Michel Foucault, *Discipline and Punish: The Birth of the Prison* (New York: Vintage Books, 1975), 177.

5. David Hurst Thomas, "Harvesting Ramona's Garden: Life in California's Mythical Mission Past," in *Columbia Consequences: The Spanish Borderlands in Pan-American Perspective,* vol. 3, ed. David Hurst Thomas (Washington, D.C.: Smithsonian Institution Press, 1991), 135.

6. Lisbeth Haas, *Conquest and Historical Identities in California, 1769-1936* (Berkeley: University of California Press, 1995), 33-7.

7. Leonard Pitt and Dale Pitt, *Los Angeles A to Z* (Berkeley: University of California Press, 1997), 14.

8. Tomás Ybarra-Frausto, "The Chicano Arts Movement," in *Exhibiting Cultures: The Poetics and Politics of Museum Display,* eds. I. Karp and S. D. Lavine (Washington, D.C.: Smithsonian Institute Press, 1991), 137.

9. Ibid.

10. Ibid.

11. Luis Aponte-Parés, "Casitas: Place and Culture," *Places* 11, no. 1 (winter 1997): 54, 56.

12. Dream Houses Exhibition, Hostos Arts Gallery, Bronx, 1998. Nina Rappaport, Curator. Exhibiting the works of architects Warren A. James, Miguel Rivera, and Madeleine Sánchez.

13. Frances R. Aparicio and Susana Chávez-Silverman, eds., *Tropicalizations: Transcultural Representations of Latinidad* (London: University Press of New England, 1997), 19.

14. Tomás Ybarra-Frausto, "The Chicano Arts Movement," 133.

15. Guillermo Gómez-Peña, *New World Borders* (San Francisco: City Lights, 1996), 5.

# Index

# About the Contributors

**Lalo Alcaraz** draws "L.A. Cucaracha," an editorial cartoon based in the *LA Weekly* and syndicated in several newspapers and magazines in the United States and abroad. He began drawing editorial cartoons in 1985 for his college paper, the *Daily Aztec* at San Diego State University, where he earned a B.A. in art and environmental design. He also has an M.A. in architecture from the University of California, Berkeley. He was born across the San Diego-Tijuana border just in time to acquire U.S. citizenship in 1964. He was a staff writer on the Fox TV show *Culture Clash*, coedits the satirical magazine *POCHO*, and performs in the sketch comedy troupe POCHO! His animation division, Animaquiladora, has created award-winning computer animation that has been screened around the United States and in Japan. POCHO's web site, the Virtual Varrio, is at *www.pocho.com*.

**Luis Alfaro** is a queer Chicano performance artist who writes poetry, essays, and drama. He is codirector of the Latino Theater Initiative at the Mark Taper Forum. He has a CD of spoken word performance titled *Downtown* and is the recipient of a MacArthur Foundation Fellowship.

**Gloria Enedina Alvarez,** Chicana, poet-intermedia artist, literary translator, and curator, presently teaches creative writing in public schools, universities, libraries, and art centers. She has received grants from the California Arts Council, the National Endowment for the Arts, and the City of Los Angeles Cultural Affairs Department. She has published and read widely in the United

States and Latin America. Her books of poetry in English and Spanish include *La excusa/The Excuse* and *Emerging en un mar de olanes.*

**Laura Alvarez** received her B.A. in fine arts from the University of California, Santa Cruz, spending her senior year at the University of Leeds in England. She received her M.F.A. in painting from the San Francisco Art Institute. In addition to working in visual media, she uses writing and music to further explore the issues that interest her.

**Carlos Avila** is a Los Angeles-based filmmaker who earned his M.F.A. at UCLA's School of Theater, Film, and Television. He is the recipient of a 1995-1996 Rockefeller Foundation Intercultural Fellowship. His award-winning films *Distant Water* (1991) and *La Carpa* (1993) have been presented at the Sundance Film Festival, the Seattle International Film Festival, and the FFICS Festival in Tokyo.

**Theresa Chavez** has written, directed, and produced numerous interdisciplinary collaborative performance theater works. She is cofounder and artistic director of About Productions theater company. Her work has been seen at venues such as the Actors' Gang Theater, the Bilingual Foundation for the Arts, Telluride (Colorado) Theater Festival, the Mark Taper Forum Out in Front Festival, the Public Theater (New York) New Voices Festival, the National Women's Theater Festival, and Highways Performance Space. She teaches classes in interdisciplinary studies at the California Institute of the Arts.

**Jessica Chornesky** is a freelance photographer whose work has appeared in *Time* magazine, *Spin, Elle,* and the *Los Angeles Times.* In 1995, she received a grant from Art Matter and Stanford University to document life in a Bosnian Muslim refugee camp. She is currently producing a wall calendar for 1999 featuring rock stars from Latin American countries.

**Margaret Crawford** is Chair of the History and Theory Program at the Southern California Institute of Architecture. She was educated at the University of California and the Architectural Association in London and received her Ph.D. from the Graduate School of Architecture and Urban Planning at UCLA. Her research is focused on the twentieth-century American built environment. She edited *The Car and the City: The Automobile, the Built Environment, and Daily Life in Los Angeles* (1990) and the forthcoming *Everyday Urbanism* and is the author of *Building the Workingman's Paradise: The Design of American Company Towns* (1995). She has also written numerous articles on company towns, housing, and urban design. She is currently working on a book on informal public spaces in Los Angeles.

**Michael Dear** is Director of the Southern California Studies Center and professor of geography at the University of Southern California.

**Ulises Diaz** (Southern California Institute of Architecture, 1992) is a partner in ADOBE LA and Studio Dos ó Tres. He studies the urban landscape culture and teaches architectural and three-dimensional workshops for youth in which he addresses issues of public open space in Los Angeles. He has exhibited at the Museum of Contemporary Art in Los Angeles, Wexner Center in Columbus, and the Gamel Dok Architecture Museum in Copenhagen. He is currently adjunct faculty at SCI-Arc and Woodbury University.

**Yvette C. Doss** is editor and copublisher of *frontera Magazine*, an English-language Chicano/Latino music and culture magazine. Her feature articles, op-eds, and commentaries have appeared in the *Los Angeles Times*, the *Miami Herald*, the *San Jose Mercury News*, the *San Francisco Examiner*, and other daily newspapers.

**Julie Easton** has been capturing Los Angeles and its people on film for over 25 years. She has documented the city's murals, community and cultural events and has done close-up studies of L.A.'s ethnic communities. She has had a solo exhibition of her workand has exhibited in various museums, including the Boston Art Museum and the Downey Museum of Art. She has traveled and photographed extensively in Mexico, Europe, China, Southeast Asia, and North Africa. In 1998, having received a Fulbright Scholarship, she lived, photographed, and taught in Morocco.

**Christina Fernandez** is an artist-photographer and adjunct faculty at Pomona College and California State University, Long Beach. Her work examines family histories and personal stories within a social-political framework. She earned a B.A. in fine arts from UCLA in 1989 and her M.F.A. from the California Institute of the Arts in 1996. Her work has been exhibited widely in the United States and Mexico. In 1998, she was co-curator for the *Xtrascape* exhibit at the Los Angeles Municipal Part Gallery and was commissioned by the Los Angeles Center for Photographic Studies to produce a one-person exhibit to open in spring, 1999. Most recently, she completed two site-specific installations in the San Diego-Tijuana border region for InSite '97 and was guest editor for the LACPS Framework journal, "Youth Through the Lens: On Youth Arts and Culture" (December 1997).

**María Elena Fernández** is a freelance writer, poet, and performance artist. She has written about Latino popular culture for the *LA Weekly*, the *Los Angeles Times,* and *Latina* magazine. She has presented her poetry and performance art, including "Hair: Confessions of a Cha-Cha Feminist," throughout her native city as well as the Nuyorican Poets' Café. She has taught Chicano studies at Cal State Northridge and has a bachelor's degree in American studies from Yale University and a master's degree in history from UCLA.

**Harry Gamboa Jr.** is a conceptual and multidisciplinary artist who was a founding member of the Chicano vanguard art collective ASCO. He has taught film production, photography, and Chicano studies at several Southern California colleges and universities. He is the author of *Urban Exile: Collected Writing of Henry Gamboa Jr.* (1998).

**Ramón García** is a writer, scholar, and teacher. He has a B.A. in Spanish literature from the University of California, Santa Cruz, and a Ph.D. in literature from University of California, San Diego. He has published poetry and fiction in anthologies and journals such as *Story,* the *Americas Review, Best American Poetry 1996, The Paterson Review,* and *Poesida.* He has published critical work in *Aztlán.*

**Rita González** is a video artist, writer, and media curator. Her work has been screened and exhibited at the Armand Hammer Museum, the Center on Contemporary Art (Seattle), and as part of l.a. freewaves at MOCA TC and the Santa Monica Museum. She is curating a series on experimental film and video in Mexico with filmmaker Jesse Lerner. Currently, she is the Lila Wallace Curatorial Intern at the Museum of Contemporary Art in San Diego.

**Robert Alexander González,** architect, is founder and editor of the journal *AULA: Architecture and Urbanism in Las Américas* and assistant editor of *Places: A Forum of Environmental Design.* He has taught at Arizona State University, Woodbury University, and Universityof California, Berkeley, where he is currently pursuing a Ph.D. in architectural history, specializing in Latin American and borderland architecture and urbanism.. Originally from the border city of Laredo, Texas, he received a B.Arch. from the University of Texas at Austin and a S.M.Arch.S. degree from Massachusetts Institute of Technology.

**Lindsey Haley** is a Chicana-Irish poet, writer, and journalist. She was part of the third generation in her family to be born in El Paso, Texas. In 1968, when she was 9, her family relocated to Venice, California, where she still lives today. Her work has appeared in *La Opinion, LA Weekly, Belvedere Citizen-Eastside Journal, Lowrider Magazine, Public Art Review,* and *Frame-Work.* She has presented her work throughout the Southwest and recently in Mexico City. She is working on a collection of short stories.

**Anthony Hernández** is a self-taught photographer raised in Los Angeles. His photographs, both black-and-white and color, have been purchased by major museums in the United States and Europe. He has received numerous awards for his work, including three National Endowment for the Arts Fellowships (1975, 1978, 1980), the Charles Pratt Memorial Award (1993), the DG Bank-Forderpreis Fotografie Award (Sprengel Museum, Hannover, Germany, 1995), the Higashikawa Prize (1998), and the Rome Prize Fellowship (1998-1999).

**Gustavo Leclerc** is a designer, artist, and founding member of ADOBE LA, where he is project manager of Laboratorio de Experimentación Urbana and Bordergraphies-Huellas Fronterizas: Retranslating the Urban Text in Los Angeles and Tijuana. He received a degree in architecture from the Universidad Veracruzana, Mexico, and has extensive experience in designing architectural projects and in teaching. He is on the SCI-Arc design faculty and has lectured throughout the United States on architecture, design, art, and cultural criticism. He was a Loeb Fellow at Harvard University in 1998-1999.

**Jesse Lerner** is a documentary film and video maker based in Los Angeles. His short film *Natives* (1991, with Scott Sterling) and the feature-length experimental documentary *Frontierland* (1995, with Rubén Ortiz-Torres) have won numerous prizes at film festivals in the United States, Latin America, and Japan. His critical essays on photography, film, and video have appeared in *Afterimage, History of Photography,* and other media art journals.

**John A. Loomis,** architect, critic, and historian, is Chair of Architecture at the California College of Arts and Crafts in San Francisco. He is the author of *Revolution of Forms: Cuba's Forgotten Art Schools* (1999).

**Alma López** is a visual and public artist born in Sinaloa, Mexico, and raised in Los Angeles, California. She has created, painted, and designed murals and digital murals in collaboration with artists and community members of California, Nevada, Texas, and Wisconsin. She was awarded a 1997 Design Excellence in Public Art from the City of Los Angeles Cultural Affairs Department and a 1998 COLA (City of Los Angeles) Individual Artist Grant. She received her M.F.A. degree from the University of California, Irvine. She is director and cofounder of Homegirl Productions, a public and visual art collaborative that focuses on the chiasmus of African American and Chicano/a experiences.

**Rogelio Villarreal Macías** is a writer and a cultural critic living in Mexico City. He is the publisher and editor of *La Pos Moderna,* an alternative magazine focusing on postmodern popular art and culture in Mexico City. He has written extensively on issues of Mexican urban identity and the politics of cultural representation.

**Rubén Martínez** is a poet, journalist, and performer. He is an associate editor at Pacific News Service and the author of *The Other Side: Notes from the New L.A., Mexico City and Beyond* (1993). A second book, on migrant culture in the U.S-Mexican borderlands, is forthcoming in 1999.

**Pedro Meyer** is a photographer and writer. He is the founder and editor of *ZoneZero,* an alternative online interactive website that functions as an exhibition space, critical forum, and electronic journal.

**Don Normark** began his fifty-year career in photography with his 1948 and 1949 photographs from Chávez Ravine, and after 10,000 published magazine photos, he has returned to them. He is working on a film based on these images, and a book of photographs and interviews, *Chavez Ravine/A Los Angeles Story,* is forthcoming.

**Alessandra Moctezuma** is an artist and founding member of ADOBE LA. In 1994, she received an MFA from UCLA, and she is a recipient of various awards, including the UCLA Arts Council Award and the Carlos Almaraz Scholarship. She has participated in group exhibitions and performances, lectured widely on issues of art, Latino culture, and vernacular architecture, and produced the videos *Andale, Andale, Frida,* and *Latino Urban Revisions.*

**Ramona Ortega** is a roaming freelance journalist currently living in the Echo Park neighborhood of Los Angeles.

**Rubén Ortiz-Torres** was born in Mexico City and lives and works in Los Angeles and Mexico City. He studied fine arts at the Escuela Nacional de Artes Plásticas, Mexico City, and architecture at Harvard University. He was a Fulbright scholar at the California Institute of the Arts in Valencia, California. He is a cross-border artist who works in a variety of media, including painting, photography, mixed media, and film. He has exhibited extensively in Mexico, the United States, Europe, Australia, and Brazil. His films include *How to Read Macho Mouse* (1993) and *Frontierland* (1995), a collaboration with Jesse Lerner.

**Reynaldo Rivera** is a Los Angeles-based Mexican photographer who, for the last two years, has been capturing the psyche of Los Angeles in bold images in a series titled "Oblivion Seekers."

**James Rojas,** a native of East L.A., received a master's degree in urban planning from the Massachusetts Institute of Technology. His research onthe ways Latinos use public space to define their community has been cited in various academic texts, and he has presented it at numerous conferences. He spent three years in the Peace Corps in Hungary. His current work at the Los Angeles County Metropolitan Transportation Authority focuses on how pedestrians react to and utilize transit shelters.

**Camilo José Vergara** received his M.A. in sociology at Columbia University. His photographs documenting the physical and cultural transformations of the central city in American metropolises have been internationally exhibited and are held in several major photographic archives. His most recent book is *The New American Ghetto.*

**Raúl Villa** is Associate Professor of English and American Studies at Occidental College in Los Angeles. His interests in issues of urbanization and Latino cultural expression are taken up in his forthcoming book, *Barrio-Logos: The Dialectic of Space and Place in Urban Chicano Culture.*

**Jennifer Wolch** is Professor of Geography and Urban Planning at the University of Southern California.

**Tomás Ybarra-Frausto** is Associate Director for Arts and Humanities at the Rockefeller Foundation. His projects with the division include the Humanities Residency Fellowship Program, Museum Program, U.S. Mexico Fund for Culture, and La Red Latinoamericana de Productores Culturales. Prior to his position with the Rockefeller Foundation, he was a tenured professor at Stanford University in the department of Spanish and Portuguese. He has served as chair of the board of the Mexican Museum in San Francisco, serves as a chair of the Smithsonian Council, and has written and published extensively, focusing primarily on Latin American and U.S. Latino cultural issues.

LATINO CENTER
Tufts University
226 College Avenue
Medford, MA 02155